Lecture Notes of the Institute for Computer Sciences, Social Informatics and Telecommunications Engineering   564

The LNICST series publishes ICST's conferences, symposia and workshops.

LNICST reports state-of-the-art results in areas related to the scope of the Institute. The type of material published includes

- Proceedings (published in time for the respective event)
- Other edited monographs (such as project reports or invited volumes)

LNICST topics span the following areas:

- General Computer Science
- E-Economy
- E-Medicine
- Knowledge Management
- Multimedia
- Operations, Management and Policy
- Social Informatics
- Systems

Anthony L. Brooks
Editor

ArtsIT, Interactivity and Game Creation

12th EAI International Conference, ArtsIT 2023
São Paulo, Brazil, November 27–29, 2023
Proceedings, Part I

 Springer

Editor
Anthony L. Brooks 🅳
Aalborg University
Aalborg, Denmark

ISSN 1867-8211   ISSN 1867-822X (electronic)
Lecture Notes of the Institute for Computer Sciences, Social Informatics
and Telecommunications Engineering
ISBN 978-3-031-55318-9   ISBN 978-3-031-55319-6 (eBook)
https://doi.org/10.1007/978-3-031-55319-6

This Springer imprint is published by the registered company Springer Nature Switzerland AG
The registered company address is: Gewerbestrasse 11, 6330 Cham, Switzerland

Paper in this product is recyclable.

Preface

I am delighted to introduce the proceedings of the twelfth edition of the European Alliance for Innovation (EAI) International Conference on Arts and Technologies, Interactivity, and Game Creation (ArtsIT 2023). This conference brought together researchers, developers and practitioners around the world who are leveraging and developing topics associated with IT in the Arts.

The technical program of ArtsIT 2023 consisted of 40 full papers presented in face-to-face or remote online sessions. Special track sessions at the event matched the groupings in these proceedings, namely 'Exploring new frontiers in Music Therapy' – Special Track (4 papers); 'Network Dance & Technology' – Special Track (4 papers); and 'Computational Art and the Creative Process' – Special Track (6 papers). Other papers are from the main proceedings tracks, which are grouped herein under 'Alternative realities, immersion experiences, and arts-based research' (7 papers); 'Games' (4 papers); 'Interactive technologies, multimedia, and musical art' (8 papers); and 'Human at Centre' (7 papers).

Coordination with the conference managers was essential for the success of the conference and the whole team acknowledge and appreciate their constant support and guidance. From my side, it was also a great pleasure to work with such an excellent organizing committee team and I acknowledge their hard work in organizing and supporting the conference.

Anthony L. Brooks

Editorial

The 12th EAI International Conference: ArtsIT, Interactivity & Game Creation (ArtsIT 2023) was hosted November 27th – 29th at the Institute of Philosophy and Human Sciences – IFCH – Marielle Franco and Fausto Castilho Auditoriums, UNICAMP, the State University of Campinas, São Paulo, in Brazil.

The ArtsIT conference once again brought together a wide array of cross-/inter-/trans-disciplinary researchers, practitioners, artists, and academia to present and discuss the symbiosis between art (in its widest definition) and information technology applied in a variety of distinct fields. Since 2009 ArtsIT has been steered[1] towards becoming acknowledged as a high quality scholarly leading scientific forum for the dissemination of cutting-edge research results in the intersection between art, science, culture, performing arts, media, and technology. The role of artistic practice using digital media also serves as a tool for analysis and critical reflection on how technologies influence our lives, culture, and society. ArtsIT is therefore considered not only as a place to discuss technological progress but also a place to reflect on the impact of art and technology on sustainability, responsibility, and human dignity.

Campinas is a Brazilian city located in the southeast region of São Paulo State. It is situated a few kilometres from the Tropic of Capricorn, which, at about 23°27' S, passes through the region north of São Paulo, roughly marking the boundary between the tropical and temperate areas of South America. São Paulo (SP) is the first Brazilian city to become a metropolis without being a capital, exerting significant national influence, and, because of its elevation, it experiences a temperate climate. The city is home to several universities, including Unicamp University, which ranks in the top 15% of the best universities globally, and among the top 3 in Latin America, which as the State University of Campinas, in its 57 years of existence has achieved international recognition through its building and disseminating of knowledge and innovation.

ArtsIT 2023 was hosted under Unicamp by the Interdisciplinary Nucleus of Sound Studies (NICS), as part of its 40th anniversary. Formed in 1983, NICS' primary objective has been to research different manifestations of sound, in all its aspects and applications, as a source of informational, cognitive, and creative content. The nucleus was founded with a vocation for interdisciplinary research, addressing contemporary innovative themes that ongoingly are updated and include Analysis of Musical Practices and Sound Art Studies; Interactive Environments and Interfaces in Extended Music; Analysis, Synthesis, and Perception of Sound Scenes; Sound pollution; Analysis of sounds in Brazilian forests, phonetics of their fauna and native populations; electronic music; music therapy, and ethnomusicological records (amongst others). Such themes associated well with the hosting of ArtsIT as is evident by the scope of submissions call for texts – see at the web site https://artsit.eai-conferences.org/2023/ under URL menu item "For Authors/Calls".

---

[1] Since 2009 the (near) annual international conference ArtsIT has been steered by Anthony Lewis Brooks.

Further evidence of alignment between ArtsIT and NICS extended beyond the main track call with the offering of special tracks on eight specialist subjects. These were, namely, special track on:- *'Affective and Emphatic Physical and/or Digital Experiences to Human Body'* - chaired by Dr. Rachel Zuanon (Arts Institute and School of Civil Engineering, Architecture and Urban Design, State University of Campinas [UNICAMP]); *'Amplifying Creativity: Exploring Musical Interfaces – An Artistic and Historical Approach'* - chaired by Elena Partesotti (NICS Lab, Unicamp University, Brazil) and Marcelo Wanderley (Music Research – IDMIL – McGill University, Canada); *'Artificially Generated Content: Identification and Ownership Protection in the Age of AI'* – chaired by Paula Dornhofer Paro Costa (Dept. of Computer Engineering and Automation, School of Electrical and Computer Engineering, University of Campinas, Brazil); *'Computational Art and the Creative Process: Exploring Human-Computer Co-creation'* – co-chaired by Artemis Moroni (DISCF – CTI Renato Archer, Brazil), Jônatas Manzolli (IA/NICS – UNICAMP University, Brazil), Elena Partesotti (NICS – UNICAMP University, Brazil), and Manuel Falleiros (NICS – UNICAMP University, Brazil); *'Music Therapy, Technology, Embodied Cognition and Rehabilitation: Exploring New Frontiers'* – co-chaired by Elena Partesotti (NICS Lab, Unicamp University, Brazil) and Wendy Magee (Music Therapy Program, Temple University, USA); *'Expression in Human-Robot Interaction'* – co-chaired by Murillo Rehder Batista and Artemis Moroni (both from CTI Renato Archer, Brazil); *'Network Dance & Technology'* – chaired by Daniela Gatti (Corporal arts Department – Arts Institute, Unicamp University, Brazil); and a multi-language (English, Portuguese, Spanish or Italian text submission) offering of a *'Special Track on Musical Meanings in Ubimus/Significados Musicais em Ubimus'* – co-chaired by Damián Keller and Luzilei Aliel (both from NAP, Universidade Federal do Acre, Universidade Federal da Paraíba, Brazil) and Micael Antunes (Universidade de Campinas, Brazil). Each of these eight special tracks offered authors opportunities for distinct specialized submissions that were handled by luminary chairpersons alongside selected panels of experts in the subject field as track committee members. Each special track had its own web presence under ArtsIT 2023 with each URL site (page) detailing with abstracts/calls on the distinct speciality on offer.

Additionally, each chair and aligned special track technical program committee (TPC) were offered to conduct their own specialized physical session at the actual event where the chair and TPC could lead and manage the session as a self-contained 'sub-event'. Thus, offering new opportunities for young career academics in the specialized fields to meet pioneering leaders, luminaries in their field, and peers for networking and insightful debate and discussions at greater depth than in a more general session. Groupings of the special tracks is also reflected in these proceedings from the ArtsIT 2023 event. A posters session supported the main and special tracks.

The following text offers insight to the actual physical 'happenings' that were incorporated under ArtsIT 2023 to enhance the delegates' experiences.

Having a commitment to diversity and inclusivity, the local organizing team gave space at the venue to the native indigenous community whose representative Brazilian artisans realized handmade nametags for delegates. Such representatives, who have connections with UNICAMP University, were present during the first two days of the conference offering typical indigenous body painting, thus providing delegates a unique opportunity to engage with and appreciate their rich cultural heritage. This artistic choice aimed to contribute to a deeper and more inclusive intercultural dialogue, promoting an understanding and appreciation of the cultural roots that shape the richness and diversity of Brazilian society, as well as the political voice of indigenous communities in the pursuit of a more just and equitable society. An outcome from this understanding and appreciation was ever present each day in the form of the indigenous logo of ArtsIT 2023 – realized as a manifestation inspired by the intersection of art and technology, as shown in image two of the web page automatic 'slider' (see above link) and herein shared below as Fig. 1.

Fig. 1. Brazilian indigenous logo of ArtsIT 2023

On a special note, on the first day after the Technical Sessions, a virtual reality experience was shared at the CPV Hotel by Professor Ivani Santana, Artemis Moroni, Felipe Mammoli, and their team who presented 'GaiaSenses', showcasing audio-visual compositions created using data from planetary platforms (see https://www1.cti.gov.br/sites/default/files/jicc-2021-paper-7.pdf).

ArtsIT 2023 featured four luminary plenary keynote speakers who were each offered the opportunity to publish individually in this proceedings, however, none were forthcoming.

The first keynote talk, on the 27th of November opening morning, was by Wendy Magee who is professor under the Music Therapy Program at Temple University, USA. Magee is also on the Editorial Board of the Journal of Music Therapy, and board member of the International Association of Music and Medicine. The title of the keynote was *'Access, agency and aesthetics: developments music therapy and technology'*.

The second ArtsIT 2023 keynote talk was given on the afternoon of 27th of November by Marcelo Wanderley who is Professor of Music Technology at McGill University, in Montreal, Canada. Wanderley's research interests include the design and evaluation of digital musical instruments and the analysis of performer movements. He co-edited the electronic book "Trends in Gestural Control of Music" in 2000, co-authored the textbook "New Digital Musical Instruments: Control and Interaction Beyond the Keyboard" in 2006, and chaired the 2003 International Conference on New Interfaces for Musical Expression (NIME03). He is a member of the Computer Music Journal's Editorial Advisory Board and a senior member of the ACM and the IEEE. The title of the keynote was '*Five Decades of Computer Music Interfaces: from ICMC & CMJ to NIME*'.

The third ArtsIT 2023 keynote talk, given on the morning of 28th of November, was by Dr Frédéric Bevilacqua, Head of the Sound Music Movement Interaction team at IRCAM in Paris. Bevilacqua is co-founder of the International Conference on Movement and Computing and is part of the joint research lab on Science & Technology for Music and Sound (between IRCAM, CNRS, and Sorbonne Université), a collaboration focusing upon the modelling and design of interaction between human movement, sound, and the development of gesture-based digital musical instruments. The title of the keynote was '*Sound-Music-Movement Interaction: from listening to performing, from general public to musician, from solo to collective experiences*'.

The final ArtsIT 2023 keynote was given by Professor Anderson Rocha on the event's closing morning on the 29th of November. Rocha is full professor of Artificial Intelligence and Digital Forensics at the Institute of Computing, University of Campinas (Unicamp), Brazil. He is the Director of the Artificial Intelligence Lab (Recod.ai) and Institute Director for the 2019–2023 period. The title of Rocha's keynote was '*How to Live with Synthetic Realities: ChatGPT, Midjourney, Dall-E2, Stable Diffusion, and others*'.

More details on the keynotes and the speakers' profiles are available at the conference web site. Also available at the web site is the full program of delegates' presentations aligned to the texts within this proceedings.

Extra to the proceedings was an optional excursion for delegates to Sirius, the new Brazilian synchrotron light source hosted at the Brazilian Synchrotron Light Laboratory (LNLS) on 30th November – see https://lnls.cnpem.br/sirius-en/.

There were two best papers awarded at ArtsIT 2023 and heartfelt congratulations are shared in acknowledging the authors who each received a gratis registration to ArtsIT 2024 (UAE). These prize-winning papers are herein listed and are available for reading within this volume:

(1) *Art as an expanded field: the case of the r/place social experiment*. Authors: Marcela Jatene Cavalcante Botelho and Hosana Celeste Oliveira (both from The Federal University of Pará [UFPA])

(2) *Fostering Collaboration in Science: Designing an Exploratory Time Travel Visualization*. Authors: Bruno Azevedo (Centro ALGORITMI, EngageLab – University of Minho); Francisco Cunha (University of Minho); Pedro Branco (University of Minho, Dep. of Information Systems [DSI])

Overviewing the ArtsIT event is to once again emphasize that it has a rich history and conjoined is its growing community of global scholars who are acknowledged as leading within their respective fields.

Previous editions of ArtsIT have been hosted at the following locations: ArtsIT 2022 was hosted by The Centro de Investigação em Artes e Comunicação (CIAC) under The University of Algarve, in Faro, Portugal. In 2021, ArtsIT was targeted to be hosted as a hybrid event at the UNESCO Creative City of Media Arts Karlsruhe, Germany, and in Cyberspace, however, it had to be changed to a fully-fledged online conference due to the COVID-19/Coronavirus pandemic. ArtsIT 2020 was also hosted as a fully-fledged online conference due to the pandemic. In 2019, ArtsIT was hosted physically at CREATE (The Institute for Architecture, Design and Media Technology) under Aalborg University main campus, in Aalborg Denmark. In 2018 the University of Minho in Braga, Portugal hosted ArtsIT, and in 2017 The Technological Educational Institute of Crete in its capital city of Heraklion, Greece hosted the ArtsIT event. ArtsIT 2016 was again hosted at CREATE (The Institute for Architecture, Design and Media Technology) but this time at Aalborg University's Esbjerg campus, Denmark, where CREATE's foundational Medialogy education was originated and established. ArtsIT was not run in 2015. In 2014 the ArtsIT international conference was hosted in Turkey, at the historic Minerva Han, the Communications Centre of Sabancı University, located in Karaköy in the very heart of Istanbul city. ArtsIT 2013 was hosted under the Department of Informatics, Systems and Communication (DISCo), University of Milano-Bicocca in Milan, Italy and in 2011, ArtsIT was again hosted under CREATE at Aalborg University's Esbjerg city campus in Denmark. The inaugural ArtsIT was held in 2009 at Yi-Lan, Taiwan. Hyperlinks to all prior ArtsIT event web sites with details of hosts, committees, etc., are found on the latest ArtsIT front web page (scroll down to view on lower right side).

Acknowledgements

In closing this editorial text of the ArtsIT 2023 proceedings, acknowledgement is stated of the associated EAI team and especially the conference managers Veronika Kissova and Radka Vasileiadis. Distinctive thanks are of course sent to the local Brazilian team who worked so hard to organize the ArtsIT 2023 hosting, these being: local organizing chair Elena Partesotti (NICS Lab, Unicamp University, Brazil); general co-chairs, Manuel Falleiros (NICS, Unicamp University, Brazil) and Artemis Moroni (ICT Renato Archer, Brazil); technical program committee chairs, Marcelo Wanderley (Music Research – IDMIL – McGill University, Canada) and Wendy Magee (Music Therapy Program, Temple University, USA); and the many others as listed at the conference web site under committees. Recognition also to Dante Pezzin, Edelson Costantino and Professor Dr. José Fornari for their commitment to their roles in supporting and realizing the event. Additionally, Rafael Brandão and the COCEN team are acknowledged for their efforts in advertising the event. Gratitude to Dr. Josué Ramos from CTI Renato Archer for supporting this event, as well as to Murillo Batista and Cleide Elizeu da Silva for the extensive promotion on social media. Also, thanks to Bruno Azevedo for his support in the physical running of the event in sessions alongside the local organizing chair. Special gratitude is also extended to the Brazilian Synchrotron Light Laboratory (LNLS), especially Gustavo Moreno, who made the excursion to Sirius possible.

Behind the scenes are the 'technical crew' who are often hidden to attendees; thus, it is important to recognise their contribution and support. Heartfelt thanks therefore are extended to Dante Pezzin, Letícia Zima, Edelson Costantino, Ingryd Sousa, Jeovane Lima, Edson Pfützenreuter, Eduardo Goldenberg, Kimberly Oliveira, Caroline Okuyama, Vinícius Lima, Aden Moreira, Franco Simões, Ingryd Sousa, Guilherme Zanchetta, Ricardo Vieira Cioldin, Márcio Massamitsu Ota, Valério Freire Paiva, José Maria Otávio, - these the crew who worked tirelessly behind the scenes to ensure the seamless execution of the conference. Their expertise and dedication were instrumental in overcoming any technical challenges, contributing to the overall success of the event and hopefully no name is missed from this valuable involvement.

The invaluable and important contributions of ArtsIT 2023 student volunteers deserve special recognition due to their enthusiasm, commitment, and proactive involvement that significantly enhanced the conference atmosphere… ArtsIT ongoingly posits its mission statement to support students in their gaining positive experiences from attending such high quality international conferences as offered by the ArtsIT series of editions, where volunteering can influence, inspire and motivate becoming themselves authors and presenters that can be life-changing vocationally and otherwise…. whilst names are not herein listed, 'you know who you are': Thanks and good luck in your future endeavours, we hope to see you again at a future ArtsIT event.

For those whose poster, very short paper, and papers from the Ubimus Special Track couldn't be published due to EAI/Springer guidelines but were showcased during the event, the efforts are recognised herein alongside a mention of the importance of contributions to the conference's vibrant atmosphere: These were namely: Douglas Bazo de Castro for his Poster presentation; Camila Gonçalves, Rafael S. Oliveira, Audrey T. Tsunoda, Percy Nohamafi for their very short paper presentation; Ivan Simurra, Damián Keller, Celio Marcos and Marcello Messina for their presentation of the papers related to Ubimus' Special Track.

Finally, it is important to state that everyone involved with the growing ArtsIT community of scholars and its (near) annual hosting look forward to welcoming, if possible, all readers at one of its upcoming editions. Attendance as either an audience member, as a passive delegate attendee, or even (ideally) as an active future presenter at an ArtsIT event to share your own research within the scope of fields covered by the title *Arts and Technologies, Interactivity, and Game Creation*, which between November 13th to 15th 2024, in its 13th edition, will be hosted at New York University in Abu Dhabi, United Arab Emirates: see https://artsit.eai-conferences.org/2024/ - we all hope to meet and greet you there and meanwhile wish you good health, happiness, and well-being – until we meet again….

The ArtsIT 2023 team.

(Editorial text composed and edited by Anthony Brooks, with important contributions from: Elena Partesotti, Artemis Moroni, Manuel Falleiros, Bruno Azevedo, and Leticia Zima).

Organization

Steering Committee

Anthony Brooks   Aalborg University, Denmark
Imrich Chlamtac   Bruno Kessler Professor, University of Trento, Italy

Organizing Committee

General Chair

Elena Partesotti   Unicamp, Brazil

General Co-chairs

Manuel Falleiros   Unicamp, Brazil
Artemis Moroni   CTI Renato Archer, Brazil

TPC Chairs

Marcelo Wanderley   McGill University, Canada
Wendy Magee   Temple University, USA

TPC Co-chairs

Jin Hyun Kim   University of the Arts, Helsinki, Finland
Luca Turchet   Trento University, Italy

Sponsorship and Exhibits Chair

Dante Pezzin   Unicamp, Brazil

Local Chairs

Dante Pezzin   Unicamp, Brazil
Cleide Elizeu da Silva   CTI Renato Archer, Brazil

Publicity and Social Media Chair

Guilherme Zanchetta   Unicamp, Brazil

Publications Chair

José Eduardo Fornari Novo Junior   Unicamp, Brazil

Web Chairs

Rodolfo Luis Tonoli   Unicamp, Brazil
Gustavo Araújo Morais   Unicamp, Brazil

Posters and PhD Track Chair

Bruno Azevedo Minho   Universidade do Minho, Portugal

Panels Chairs

Elena Partesotti   Unicamp, Brazil
Manuel Falleiros   Unicamp, Brazil
Artemis Moroni   CTI Renato Archer, Brazil
Bruno Azevedo   Universidade do Minho, Portugal

Technical Program Committee

Adriano Claro Monteiro   Federal University of Goiás, Brazil
Alex Street Anglia   Ruskin University, UK
Alexandre Zamith Almeida   Unicamp, Brazil
Alfredo Raglio   University of Pavia, Italy
Anésio Azevedo Costa Neto   Federal Institute of Education, Science and
                                    Technology of São Paulo, Brazil
Antonio Rodà   University of Padova, Italy
Artemis Moroni   CTI Renato Archer, Brazil
Atau Tanaka   Goldsmiths, University of London, UK
Bruno Azevedo   Universidade do Minho, Portugal
Camila Acosta Gonçalves   Certified Music Therapist, Brazil
Carlos Augusto Nóbrega   Federal University of Rio de Janeiro, Brazil
Carlos Mario Gómez Mejía   NAP, Federal University of Paraiba, Brazil
Claudia Núñez Pacheco   KTH Royal Institute of Technology, Sweden

Claudia Zanini	Universidade Federal de Goiás, Brazil
Damián Keller	Federal University of Acre, Brazil
Daniela Gatti	Unicamp, Brazil
David Gamella	International University of La Rioja, Spain
Eduardo Hebling	Unicamp, Brazil
Elena Partesotti	Unicamp, Brazil
Esteban Walter Gonzalez	Universidade Federal Fluminense, Brazil
Felipe Mammoli	CTI Renato Archer, Brazil
Flavio Soares Correa da Silva	University of São Paulo, Brazil
Francisco Z. de Oliveira	Unicamp, Brazil
Gabriele Trovato	Shibaura Institute of Technology, Japan
Gilberto Prado	University of São Paulo, Brazil
Giovanni De Poli	University of Padua, Italy
Hélio Azevedo	CTI Renato Archer, Brazil
Ivan Simurra	NAP, Federal University of Acre, Brazil
Ivani Santana	Federal University of Río de Janeiro, Brazil
Jan Schacher	University of the Arts Helsinki, Finland
José Eduardo Fornari Novo Junior	Unicamp, Brazil
Josué Ramos	CTI Renato Archer, Brazil
Luca Truchet	Trento University, Italy
Manuel Falleiros	Unicamp, Brazil
Marcelo Caetano	PRISM Laboratory, France
Marcello Lussana	Humboldt University, Germany
Marcello Messina	NAP, Southern Federal University, Russia
María Teresa Del Moral Marcos	Universidad Pontificia de Salamanca, AEMP, Spain
Marijke Groothuis	ArtEZ University of the Arts, Enschede, The Netherlands
Melissa Mercadal-Brotons	Escola Superior de Música de Catalunya, Spain
Murillo Batista	CTI Renato Archer, Brazil
Paula Costa	Unicamp, Brazil
Rachel Zuanon Dias	Unicamp, Brazil
Regis Rossi Alves Faria	University of São Paulo, Brazil
Ricardo Del Farra	Concordia University, Canada
Rodrigo Bonacin	CTI Renato Archer, Brazil
Rodolfo Luis Tonoli	Unicamp, Brazil
Tadeu Moraes Taffarello	Unicamp, Brazil
Tereza Raquel De Melo Alcantara Silva	Federal University of Goiás, Brazil
Tiago Fernandez Tavares	Insper, Brazil
Uwe Seifert	Universität zu Köln, Germany

Contents – Part I

Computational Art and the Creative Process

Contents – Part II

Interactive Technologies, Multimedia, and Musical Art

Human at Centre

Exploring New Frontiers in Music Therapy

Exploring the Theoretical Landscape of BehCreative: Artistic and Therapeutic Possibilities of an Extended Digital Musical Instrument

Elena Partesotti[1](✉), Gabriela Castellano[2,3], and Jônatas Manzolli[1]

[1] University of Campinas (UNICAMP), NICS, Campinas, Brazil
eparteso@unicamp.br
[2] University of Campinas (UNICAMP), Gleb Wataghin Institute of Physics, Campinas, Brazil
[3] University of Campinas (UNICAMP), Brazilian Institute of Neuroscience and Neurotechnology, Campinas, Brazil

Abstract. In the digital age, technology has become ubiquitous in various fields of knowledge, functioning as an extension of the human body – akin to a technological body. Consequently, it acquires ecological validity in our daily lives and offers a path for developing studies rooted in Embodied Cognition. This idea is illustrated by BehCreative, an Extended Digital Musical Instrument (EDMI) introduced and presented in this article. The article delineates the design, mapping and architecture of BehCreative from both musical and cognitive standpoints. Moreover, it explores how these instruments bridge the divide between traditional musical interfaces and immersive technologies, thereby redefining the limits of artistic expression and therapeutic interventions. The article highlights that BehCreative, as a hybrid instrument, holds the potential to facilitate therapeutic recovery and serve as an artistic tool for expressive purposes. Given its hybrid nature, BehCreative presents diverse possibilities for exploring users' behavioral learning, as evidenced by the results of a mentioned exploratory study. Consequently, it substantiates the hypothesis of being an instrument that warrants examination from an interdisciplinary perspective.

Keywords: Music Therapy · Embodied Cognition · Artistic Installations · Creative Empowerment · EDMI

1 Introduction

Recent research in interactive therapies, grounded in technological advancements, has highlighted the utilization of immersive experiences, including Virtual Reality (VR) [1–3]. Some of these applications are connected to the utilization of interactive musical systems and Digital Musical Instruments (DMIs) [4]. One of the reasons to consider these studies stems from the burgeoning literature and practical implementations focused on the design of DMIs in therapeutic contexts [5]. The capacity to engage patients through

A. L. Brooks (Ed.): ArtsIT 2023, LNICST 564, pp. 3–15, 2024.
https://doi.org/10.1007/978-3-031-55319-6_1

musical processes controlled by gestures holds the potential to foster constructive effects on their involvement in the therapeutic process.

This article aims to build upon prior work [4] by delving into its theoretical dimension, thereby establishing the theoretical framework for deploying Extended Digital Musical Instruments (EDMI) within music therapy and rehabilitation frameworks. Consequently, the article delves into theories of embodied cognition that bridge diverse domains encompassing technology, music therapy, art, and music. In doing so, it develops a comprehensive exploration of the interconnected relationships among these domains, presenting a novel perspective on the therapeutic and rehabilitative potential of EDMI within the music technology context. First, we define convergent aspects of DMIs [6] as prospective elements in the musical and therapeutic field. One of the most relevant points is the great potential for interactive music and immersive environments for studying human cognition and its relation to interactive musical models [7]. This discourse stems from concepts rooted in DMIs and the application of mixed reality as a framework for designing immersive therapeutic environments [8].

Further, we discuss the dialogue between our approach and the Ecological Perspective [9] as well as the paradigm of Embodied Cognition. Embodied Learning surfaces as a significant factor engaged in dialogue with the concepts of Aesthetic Resonation as posited by Swingler [10], Ellis [11], and Aesthetic Resonance [12, 13]. Subsequently, we consolidate these viewpoints to introduce the notion of EDMI and discuss a methodological approach for harnessing such extended digital instruments. This framework paves the way for developing therapeutic methods and the design of experiments within this domain.

Moreover, we also introduce the implementation of an EDMI named BehCreative. This EDMI combines virtual and physical elements, creating a hybrid environment where the boundaries between the real and virtual worlds become blurred. Beyond that, BehCreative aligns with the principles of embodied cognition, motor learning, and therapeutic modalities. In the concluding sections of the article, we contemplate the implications of constructing therapeutic processes rooted in immersive installations.

2 Theoretical Background

This session presents the main concepts supporting the approach presented here. We begin the discussion with DMIs [6]. Next, we delve into the concepts related to Mixed Reality, followed by relevant aspects of Gibson's ecological theory [9], Embodied Learning, and Musical Cognition. These various aspects are interconnected to define the notion of EDMI, a key concept that supports our research methodology and the design of an experimental setup.

2.1 Digital Musical Instruments (DMI)

Miranda & Wanderley [6] defined that DMIs employ sensing technology, computer hardware, and software to craft novel musical instruments, providing users and musicians with a diverse range of possibilities for interaction. Unlike traditional musical instruments, a DMI separates its sound production unit from the gestural control unit [6].

Consequently, any instrument or object linked to digital circuitry can generate sound as an output. DMIs exhibit four main categories: 1) alternate controllers, 2) augmented or hyper-instruments, 3) instrument-like controllers, and 4) acoustic instrument-inspired controllers [6].

2.2 Mixed Reality and the Extended Paradigm

Mixed Reality (MR) is defined as "the merging of the real environment and the virtual world" [14]. In an interactive mixed reality environment, the user experiences the space as a performer. This involves two interconnected situations: real actions (i.e., live actions performed by the user) and the processing of information, both connected in real and virtual environments [15]. The user's experience in a mixed reality performance is balanced between the real and virtual, creating a more engaging interaction. Giannachi & Benford [16] describe a mixed reality performance as "... both their mixing of the real and virtual as well as their combination of live performance and interactivity" (p. 1).

The merging perspective of the real and virtual is essential for the research presented here, as we later define the notion of Extended DMIs and introduce our studied system, BehCreative. In a mixed reality environment, digital technologies seamlessly integrate with physical interfaces in the real world, enhancing the user's musical experience. In our approach interaction is a crucial aspect of the user's experience, engaging them from both real and virtual perspectives. This characteristic offers multiple points of view for the observer, as each user explores the performative space subjectively, influenced by temporal perspective, perceptual dynamics, kinetic possibilities of their body, and the type of performance, whether continuous or discontinuous. All these points make mixed reality a highly relevant perspective, as it involves the user's performance within an immersive system that responds in real-time. BehCreative utilizes various types of sensory simulation - visual, sound, and proprioceptive - through mixed reality, allowing for exploration and study beyond the Virtual Reality (VR) perspective, thus intensifying the user's experience.

2.3 Ecology of Perception and DMIs

We can infer that one of the foundations of DMIs is grounded in the idea of how technological tools provide body extensions to the user, simplifying certain situations while amplifying the results. This is why it is important to maintain a continuous dialogue with Embodied Cognition: we need to understand how humans interact with the technological environment and how it influences our actions, such as exploration, learning, and behavior. For example, considering some perspectives from Gibson's theory of visual perception, particularly regarding artworks, he refers to artworks as ways of communication that do not contain signals to be sent to the subject, but rather direct information available to the observer, mediated by perception. As Dourish [17] summarizes:

"Gibson's starting point was to consider visual perception not as a link between optics and neural activity, but as a point of contact between the creature and its environment, an environment in which the creature moves around and within which it acts" (p. 117).

Likewise, Nöe & O'Regan [18] expound upon visual perception to elucidate the Sensorimotor Contingency Theory (SCT). According to the authors, from the perspective of the visual sensorimotor approach (i.e. which performs sensory and motor functions),

"Vision is a capability not confined to the brain alone; it encompasses the entirety of the environmentally situated perceiver... The brain's role in producing vision lies in enabling active exploration based on implicit awareness of sensorimotor contingencies" (p. 593).

The Sensorimotor Contingency Theory (SCT) says the role of movement is very important to explain visual perception and other sensorimotor contingencies. This idea aids in unraveling the dynamics within the performative space introduced in this paper. According to the ecological theory viewpoint [9], the relationship between the subject and its surrounding environment shapes human experience. Furthermore, Gibson's theory can be applied to the realm of music and technology, where the experiences of musicians and music technologies undertake the relationship between the subjects and the surrounding technological environment.

Dialoguing with Gibson's concept of affordance, we defined the notion of *Virtual Affordance* as the properties of the interactive virtual environment that suggest possible or available actions based on the user's perception. In resonance with this idea, it is possible to allow for direct manipulation of Virtual Affordances, offering degrees of freedom to the user's behavior within the performative space. These results in an exchange of information between the environment and the subject, in a cyclical process.

2.4 Art, Technology and Creative Empowerment

Our approach establishes a meaningful dialogue between Art and Technology, which holds significant importance in the clinical-therapeutic field [19]. The use of music technology as a tool for promoting prevention and treatment of psychological and physical pathologies and improving the quality of life of clients and music therapists is gaining recognition.

A significant concept related to our research is the concept of Aesthetic Resonance [12, 13] and the notion of Creative Empowerment [4]. Creative Empowerment is about empowering individuals to use technology to enhance their creativity and feel a sense of control and agency over the creative process. By providing tools and resources that allow users to express their creativity in new and innovative ways, designers can foster engagement, motivation, and satisfaction, particularly within a therapeutic process. It enables individuals to explore and develop their creative skills and talents in ways that were not previously possible.

Furthermore, interactive systems have demonstrated effectiveness in motor rehabilitation, accelerating motor recovery and facilitating the internal transfer of skills to daily life [20, 21]. In the rehabilitation field, self-expression and artistic creation play vital roles in involving and motivating clients. Our research main goal is to facilitate the performer's self-expression, as achieving Creative Empowerment is a turning point in personal and physical recovery.

These concepts should be considered within the paradigm of embedded cognition, helping readers to understand the aesthetic perspective that stems from sensorimotor and perceptual experiences. The extended instrument, manipulated through the user's performance and sense of agency, involves both cognitive and physical aspects. Manipulation, in this case, is cognitive, incorporated but also physical: a physicality that gives pleasure, and that is co-determined by Creative Empowerment.

In summary, our approach seeks to create an immersive and empowering experience for users, combining the artistic and technological aspects in a manner that enhances therapeutic outcomes.

3 Extended Digital Musical Instrument (EDMI)

After discussing the main theoretical issues related to the study presented here, we can now define the central concept of our research: the notion of Extended Digital Musical Instrument (EDMI). This concept is exemplified by the system we created, BehCreative, which serves as an instance of an EDMI.

3.1 Defining the EDMI

As described in the theoretical framework, we bring back Gibson's Theory of Visual Perception [9] to outline the notion of EDMI, as previously defined by Partesotti [22]. Gibson argued in his theory that "[w]e must perceive in order to move, but we must also move in order to perceive." This concept implies a continuum between perception and action. To understand perception, we must start from the experience of the body in the surrounding environment. Our experience is bodily mediated, and guided by sensorimotor objectives. Perception and action are intrinsically linked, guided by mirror neurons (activated by observing a person performing an action) and canonical neurons (activated in the observation of objects) [23].

The concept of EDMI goes beyond the scope of Digital Musical Instruments (DMI). Unlike DMIs, which primarily focus on digital technology and interfaces, EDMIs initially embrace the concepts of environment, ecology and immersion. These instruments prioritize corporeality as a key element in interpreting reality and shaping the overall musical experience. By emphasizing the embodied interaction between the musician and the environment, EDMIs offer a new dimension of expressive possibilities and engagement.

BehCreative, our implementation of an EDMI, reflects these principles and concepts. It is an immersive environment where we developed our research methodology and completed an exploratory study [24]. The performer's interaction in this environment is distributed throughout the performative space and depends on the real-time connection between the machines. The performer receives an immersive and visual octophonic sound response, guiding their proprioception. BehCreative becomes an extension of the performer's own body, as the user herself/himself chooses when to stop using this 'virtual' instrument mixed with their body, activating points of rest. The tool extends our abilities [22] in a dimension where perception and cognition depend on the performer's interactions within the environment.

3.2 EDMI Adaptation and Extending Functionality

Mapping is utilized within BehCreative to define potential behaviors, granting users the freedom to interact and bring them to the forefront based on their preferences. An acoustic instrument, despite allowing to be played in different ways, will always remain faithful to its physical components (material, possible frequencies, and harmonics), as well as the exploration and feedback it offers. The EDMI is a unique musical instrument that is not predefined in its functionality. Unlike acoustic counterparts, the EDMI's capabilities depend on the user's behavior, making it a versatile and immersive tool for musical expression. It embraces the concept of being extended, immersive, and open to diverse ways of interaction through the user's body. This perspective highlights the significance of embodied learning, where the user's process of engaging with the instrument involves self-reflection and body awareness, leading to Creative Empowerment. By actively involving the user's body in the musical experience, the EDMI encourages a deeper connection and understanding of the instrument's potential for creative expression.

In summary, the virtual EDMI instrument offers a dynamic and immersive musical experience that is not constrained by predefined rules. Through embodied learning and its extended and immersive features, it promotes Creative Empowerment and opens the door to intelligent and adaptive musical systems.

4 Implementation of BehCreative

After discussing the concept of EDMI, we now delve into the implementation of our system, BehCreative, focusing on three key aspects: designing, mapping, and architecture. This prototype (Fig. 1) is based on the investigation of the correspondence between color and music, such as synesthesia [22, 25].

4.1 Design of BehCreative

In the design of EDMIs, a crucial aspect is the mapping. It is a critical element for the tool's usability and effectiveness [26]. As described by [27], mapping refers to the correspondence between control parameters derived from the performer's actions or gestures and the parameters of sound synthesis. This term defines the choice of inputs within an interactive music system and during a performance, which are then transformed into expressive outputs or feedback for the performer. The selection of inputs involves analyzing and observing the collected data, which depends on the objective of the EDMI. The mapping's characteristics vary from one instrument to another; some DMIs may have intuitive mappings, correlating physical and musical gestures, while others may not be as clear to the user. The mapping holds significance as it represents the architecture of the entire technology and shapes the purpose of the interactive multimodal system. Figures 2 and 3 illustrate the distinctions between the mapping of a DMI and of an EDMI, such as BehCreative. Figure 3 depicts how BehCreative expands Miranda and Wanderley's notion of processing [6].

In fact, in BehCreative, the user becomes the performer and input into a cycle of co-determination between performance and the performative space. This input is derived

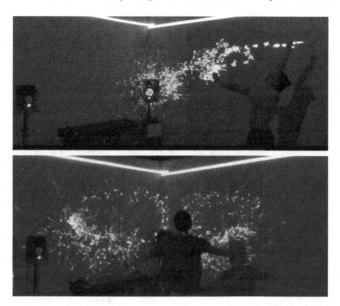

Fig. 1. Two users during an experiment with BehCreative.

from the user's own body movements and gestures in the general performative space. The data received by the MOCAP (Kinect 2) are then processed, and the body data are transformed into visual feedback (Processing Programming Environment) displayed on windows, which are projected on the three canvases around the user. Simultaneously, these data are sent via OSC to the software that interprets them to produce the sound output, including sonification and Virtual Affordances from the octophonic system (Pure Data in MAC IOS). Since the body in space is fundamental for an Extended DMI, the consideration of the user's interaction with the space in which it is immersed expands the notion of DMI proposed by Miranda & Wanderley (2006) in the input and processing of data.

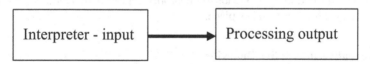

Fig. 2. Basic Diagram of a DMI proposed by Miranda and Wanderley, 2006.

4.2 Mapping of BehCreative

The mapping for an EDMI, therefore, encompasses both technical and theoretical aspects that are essential to consider in 'hybrid' applications, whether they are artistic or therapeutic.

From a technical standpoint, in BehCreative, the processing occurs through a network using the OSC protocol, involving the analysis of body tracking data obtained

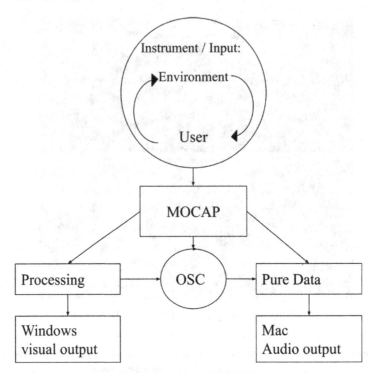

Fig. 3. Mapping of an EDMI.

from Kinect 2. This network serves two functions: it receives, processes, and shares the input data, and also acts as a general audiovisual controller. On one side, there is the Processing Programming Environment generating visuals, and subsequently, Pure Data (PD) receives the encoded movement signal, already translated into Virtual Affordances. The resulting sound is then diffused through an octophonic system, immersing the user in the BehCreative installation. This immersion is a fundamental condition of an EDMI. Through joint exploration, which leads to the proprioceptive experience of the user, the interaction with BehCreative takes place.

4.3   BehCreative Interactive Behavior

As described before, BehCreative is structured in layers, with particular attention given to sound feedback, as it functions as an extended musical instrument. Consequently, consonance and dissonance feedbacks are implemented, reacting according to the user's movements in the environment, along with various Virtual Affordances. Hence, in BehCreative, there are interactive rules that are activated through the co-determination process between the user and the environment. These interactive rules include:

Fluidity: The quantity of movement established a priori within a certain range. More balance in fluidity leads to more balanced sonification, i.e., consonance by the octophonic system. Thus, it is related to jerk.

Jerk: The variation of acceleration that influences the sense of fluidity. This variable is responsible for implementing consonance and dissonance feedback. If the subject activates the jerk, the feedback is consonant, leading to a state of fluidity. Otherwise, the auditory feedback will be dissonant, while the visual feedback is black and white.

Virtual Affordances: This represents sensorimotor feedback, which is given by the sense of proprioception - i.e., awareness of the body moving within the environment. Precise movements trigger specific sound and visual feedback, which the subject discovers through exploration and learns to use by creating their own Sensorimotor Maps.

Tunnel: This variable represents the predetermined range for positive outcomes and reflects the user's ability.

Momentum: It controls sound/silence and is linked to the number of particles shown in the visual feedback.

PoR (Point of Rest): The Point of Rest is determined by the user's movements. When the user stops or goes down on the ground, the system also stops.

5   Applications of BehCreative

5.1   BehCreative Applied in Artistic Production

It emerges that in BehCreative, the user becomes a performer, composer, and artist, producing their own unrestricted feedback. Art installations act as bridges between reality and its possible interpretations. From this perspective, BehCreative represents an interesting possibility to experience and understand the relationship between the body, technology, and art, as well as the direct communication skills that this EDMI possesses and offers.

5.2   BehCreative in Therapeutic Applications

Considering this new embodied approach, we believe it should always be applied, especially when considering DMIs for therapeutic and ludic applications through mixed reality systems.

VR can be used, among other things, to develop therapeutic and ludic applications. This is the case of Rehabilitation Gaming System (RGS), a VR based system. RGS for functional recovery of the nervous system lesions using a non-invasive approach has been developed by [28]. The key lies in providing users with a sense of control and agency over the instrument. A crucial difference between BehCreative - as Mixed Reality environment - and VR-based RGS is that they can deceive the brain for performative purposes, making the patient believe they have successfully performed a movement when the movement is limited. This stimulates the patient to improve their performance [22]. We have run the BehCreative protocol for the experiments, which included the study of the subjects' functional magnetic resonance imaging (fMRI) exam. The results obtained so far show activation of the regions connected with the rewarding system [24].

On the other hand, BehCreative takes a different perspective from RGS-based concepts, allowing the user to progress in a psychological-emotional process or motor rehabilitation through the pleasure triggered by free audiovisual feedback, experienced throughout the entire body in motion. This feedback is immediate, controlled, and pleasant. Unlike in VR, in a mixed reality system as BehCreative, the pleasure here bodily, satisfying the user because it is not just a response within a virtual game, but the body itself that is involved, providing immediate control over the feedback (i.e., the artistic production in this case). If users do not like the feedback produced, they can simply move differently to explore new paths and create their own trajectories and Virtual Affordances [22].

5.3 BehCreative in Music Therapy

Despite the development of MOCAP technologies in the artistic field, the presence of EDMI in music therapy is currently quite limited, with rare exceptions like MotionComposer [20]. Most EDMIs are primarily used in workshops or for research and are not systematically integrated into therapeutic sessions, partly due to their expensive cost. However, as we have explored in this article, the potential that this type of music technology offers to caregivers and users is vast, especially with the ability to customize the mapping to meet the client's specific needs.

For example, in BehCreative the user remains the central figure, paying attention to their movements and how they interact with the visual and sound variables. These interactions generate consonant or dissonant feedback, as well as black and white or colored outputs. BehCreative promotes bodily awareness while providing a valuable tool for investigating the behavioral learning of the user through quantitative and qualitative data recording. It captures data such as the user's jerk, sound, and visual feedback (including Virtual Affordances), and even affective variables using the Affective Sliders questionnaire [29]. The therapist can then analyze the level of user involvement based on these data if needed, as described in [30].

6 Discussion

This article underscores the potential of a research approach in the realm of musical interactivity and its related technologies, particularly in immersive therapies utilizing music and sounds for interaction. We argue that such technologies, which enhance musical control through gestures, improve engagement in the therapeutic process. The article begins by providing a comprehensive definition of Extended Digital Musical Instrument (EDMI) and then introduces the implementation of BehCreative. This interdisciplinary approach combines ecological psychology and mixed reality concepts with musical technology, primarily DMI.

The creation and implementation of BehCreative involved sound and visual environments and computational tools like Processing Programming, Pure Data or Max MXP. The experimental design anchored its methodology in embodied cognition, motor learning, and creative empowerment. Furthermore, the production of sound stimuli drew from musical composition methods adapted from sonic design for interactive environments.

BehCreative represents the convergence of these diverse elements in both its conceptual framework and practical applications.

In the studies of musical cognition and its therapeutic applications, developing new methodologies is crucial to effectively address the complexity of the area. As the field continues to evolve, interdisciplinary dialogues and collaborations with other areas of knowledge, such as philosophy, psychology, neuroscience, and technology, become increasingly important. Engaging with these fields broadens our understanding of how music is processed, experienced, and utilized in therapeutic contexts - i.e. between neuroscience and music therapy [31].

Recently, Manzolli [32] introduced Ecological Methodologies such as presented here in two aspects: 1) a network of complex data available for creative exploration, and 2) data tracking and behavior analysis in an environment where interaction with various devices indicates how the environment affects actions and derives meaning from the interaction. We foresee the potential of such methodology in the Neuroscience of Music and its application in immersive technology for music therapy.

BehCreative has recently been installed at NICS's Laboratory. As mentioned earlier, our approach aims to foster cooperation and creativity through interaction with an environment that becomes sensitive to human presence and engages in a dialogue with it [33].

Laboratories with installations such BehCreative, grounded in initiatives such as this, have tremendous potential to facilitate dialogues between various areas of knowledge within the musical field and beyond. This includes interactive music composition, musical performance with the support of new technologies, studies on music perception and musical cognition, and most importantly, applications in therapeutic domains. The concept of the Extended Digital Music Instrument (EDMI) expands beyond BehCreative's implementation and offers broad possibilities for interdisciplinary exploration.

BehCreative is an innovation resulting from interdisciplinary engagement, born from the collaboration of several research fields, including music cognition, neurotechnology, musicology, and music technology. It possesses both therapeutic and artistic applications. On one hand, it holds potential for use in rehabilitation therapy and music therapy [22]. On the other hand, it serves as an artistic installation with pedagogical potential, enabling learning about the relationship between colors and sounds, as well as the embodiment of musical concepts. As an extended instrument at a cognitive level and augmented at a mapping level, it fosters creativity in the user, empowering them to explore new creative dimensions.

7 Conclusion

In this paper, we introduce and elucidate BehCreative - an instance of our definition of an Extended Digital Music Instrument (EDMI). BehCreative a new hybrid music technology suited for both therapeutic and artistic contexts. We establish a distinction between Digital Musical Instruments (DMIs) and EDMIs. While DMIs utilize digital interfaces to amplify musicians' expressive capabilities, EDMIs seamlessly integrate gesture, sound interactions, and other sensory modalities to craft immersive and engaging musical experiences.

Our research and design process unveil that the system's behavior, aligned with the audiovisual responses of BehCreative, is intricately shapped by real-time user interactions within the installation environment. This user-initiated engagement and inherent self-organization present an innovative approach to designing artistic, therapeutic, and pedagogical musical technologies. In our pilot experiment [24] we observed neuronal activation in areas associated with the brain's reward system. The results suggest the tool's potential in stimulating neuroplasticity. Notably, even with a limited number of subjects, significant out-comes emerged.

Acknowledgement. This research was supported by São Paulo Research Foundation – Brazil (FAPESP) – Grants 2016/22619-0 and 2013/07559-3 and Brazilian National Council for Scientific and Technological Development (CNPq) – Grant 308695/2022-4.

References

1. Boeldt, D., et al.: Using virtual reality exposure therapy to enhance treatment of anxiety disorders: identifying areas of clinical adoption and potential obstacles. Front. Psych. **10**, 773 (2019). https://doi.org/10.3389/fpsyt.2019.00773
2. Grenier, S., et al.: Using virtual reality to improve the efficacy of cognitive-behavioral therapy (CBT) in the treatment of late-life anxiety: Preliminary recommendations for future research. Int. Psychogeriatr. **27**(7), 1217–1225 (2015). https://doi.org/10.1017/S1041610214002300
3. Jerdan, S.W., Grindle, M., Van Woerden, H.C., Boulos, M.N.K.: Head-mounted virtual reality and mental health: critical review of current research. JMIR Serious Games **6**(3), e9226 (2018)
4. Partesotti, E., Peñalba, A., Manzolli, J.: Digital instruments and their uses in music therapy. Nord. J. Music. Ther. **27**(5), 399–418 (2018)
5. Frid, E.: Accessible digital musical instruments—a review of musical interfaces in inclusive music practice. Multimodal Technol. Interact. **3**(3), 57 (2019)
6. Miranda, E.R., Wanderley, M.M.: New Digital Musical Instruments: Control and Interaction Beyond the Keyboard. AR Editions Inc., Middelton, US (2006)
7. Verschure, P.F., Manzolli, J.: Computational modeling of mind and music (2013)
8. i Badia, S.B., et al.: The effects of explicit and implicit interaction on user experiences in a mixed reality installation: the synthetic oracle. Teleop. Virtual Environ. **18**(4), 277–285 (2009). https://doi.org/10.1162/pres.18.4.277
9. Gibson, J.J.: The Ecological Approach to Visual Perception. Psychology Press, London (1979)
10. Swingler, T.: "That Was Me!": applications of the Soundbeam MIDI controller as a key to creative communication, learning, independence and joy (1998)
11. Ellis, P.: Developing abilities in children with special needs: a new approach. Child. Soc. **9**(4), 64–79 (1995)
12. Brooks, T., Camurri, A., Canagarajah, N., Hasselbad, S.: Interaction with shapes and sounds as a therapy for special needs and rehabilitation. In: Proceedings of the International Conference on Disability, Virtual Reality and Associated Technology, Veszprem, Hungary, ICDVRAT (2002)
13. Brooks, A.: Soundscapes: the evolution of a concept, apparatus and method where ludic engagement in virtual interactive space is a supplemental tool for therapeutic motivation. Ph.D. thesis (2011). http://vbn.aau.dk. Accessed 9 April 2022
14. Milgram, P., Kishino, F.: A taxonomy of mixed reality visual displays. IEICE Trans. Inf. Syst. **77**(12), 1321–1329 (1994)

15. Wagner, I., et al.: On the role of presence in mixed reality, **18**(4), 249–276 (2009)
16. Dourish, P.: Where the Action Is. MIT Press, London (2004)
17. Benford, S., Giannachi, G.: Performing Mixed Reality. MIT Press, Cambridge (2011)
18. Noë, A., O'Regan, J.K.: On the brain-basis of visual consciousness: a sensorimotor account. Vision and mind: selected readings in the philosophy of perception, pp. 567–598 (2002)
19. Peñalba, A., Valles, M.J., Partesotti, E., Sevillano, M.Á., Castañón, R.: Accessibility and participation in the use of an inclusive musical instrument: the case of MotionComposer. J. Music Technol. Educ. **12**(1), 79–94 (2019)
20. Holden, M.K.: Virtual environments for motor rehabilitation: review. Cyber Psychol. Behav. **8**(3), 187–219 (2005)
21. Rizzo, A.A., Kim, G.J.: A SWOT analysis of the field of virtual reality rehabilitation and therapy. **14**(2), 119–146 (2005)
22. Partesotti, E.: Extended digital music instruments to empower well-being through creativity. In: Brooks, A.L. (ed.) Creating Digitally. Shifting Boundaries: Arts and Technologies - Contemporary Applications and Concepts, vol. 241, pp. 365–401. Springer, Cham (2023). https://doi.org/10.1007/978-3-031-31360-8_13
23. Gallese, V., Fadiga, L., Fogassi, L., Rizzolatti, G.: Action recognition in the premotorcortex. Brain **119**, 593–609 (1996)
24. Partesotti, E., Feitosa, J.A., Manzolli, J., Castellano, G.: Behavioural changes and multimodal interaction within a performative space: an interdisciplinary investigation. J. Epilepsy Clin. Neurophysiol. (2020). ISSN 1676-2649. Proceedings of 6th Brain Congress, Brazilian Institute of Neuroscience and Neurotechnology
25. Partesotti, E., Tavares, T.F.: Color and emotion caused by auditory stimuli. In: ICMC (2014)
26. Iazzetta, F.: Meaning in music gesture. In: Battier, M., Wanderley, M. (eds.) Trends in Gestural Control of Music. IRCAM-Centre Pompidou, Paris, France (2000)
27. Hunt, A., Wanderley M., Kirk., R.: Towards a model for instrumental mapping in expert musical interaction. In: Proceedings of the 2000 International Computer Music Conference, pp. 209–212. International Computer Music Association, San Francisco (2000)
28. Cameirão, M.S., Badia, S.B., Duarte, E., Frisoli, A., Verschure, P.F.: The combined impact of virtual reality neurorehabilitation and its interfaces on upper extremity functional recovery in patients with chronic stroke. Stroke **43**(10), 2720–2728 (2012). https://doi.org/10.1161/STROKEAHA.112.653196. Epub 7 Aug 2012. PMID: 22871683
29. Betella, A., Verschure, P.F.: The affective slider: a digital self-assessment scale for the measurement of human emotions. PLoS ONE **11**(2), e0148037 (2016)
30. Partesotti, E., et al.: Analysis of affective behavior in the artistic installation moviescape. In: Brooks, A.L. (ed.) ArtsIT 2022. LNCS, vol. 479, pp. 327–345. Springer, Cham (2022). https://doi.org/10.1007/978-3-031-28993-4_23
31. Koelsch, S.: Investigation emotion with music: neuroscientific approaches. Ann. N. Y. Acad. Sci. **1060**, 412–418 (2005). https://doi.org/10.1196/annals.1360.034
32. Manzolli, J.: A auto-organização como agente do processo criativo e laboratório de significação musical. Acta Scientiarum **44**(1), e62240 (2022)
33. Wassermann, K.C., Eng, K., Verschure, P.F.M.J., Manzolli, J.: Live soundscape composition based on synthetic emotions. IEEE Multimedia **10**(4), 82–90 (2003). https://doi.org/10.1109/MMUL.2003.1237553

Adapting the Emobook Life Story Book App for Reminiscence Focused Music Therapy in Dementia Care: An Interdisciplinary Participatory Design Approach

Noelia Gerbaudo-Gonzalez[1]([⊠]) [iD], Alejandro Catala[2,3] [iD],
Nelly Condori-Fernandez[2,3] [iD], and Manuel Gandoy-Crego[4] [iD]

[1] Departamento de Psicoloxía Evolutiva, Universidade de
Santiago de Compostela, Santiago de Compostela, Spain
noelia.gerbaudo@usc.es

[2] Centro Singular de Investigación en Tecnoloxías Intelixentes (CiTIUS),
Universidade de Santiago de Compostela, Santiago de Compostela, Spain
{alejandro.catala,n.condori.fernandez}@usc.es

[3] Departamento de Electrónica e Computación, Universidade de Santiago de Compostela,
Santiago de Compostela, Spain

[4] Departamento de Psiquiatría, Radioloxía, Saúde Pública, Enfermaría e Medicina,
Universidade de Santiago de Compostela, Santiago de Compostela, Spain
manuel.gandoy@usc.es

Abstract. Life Story books are frequently employed to facilitate reminiscence interventions, but their use in music therapy remains limited in the scientific literature. There is a paucity of research detailing the design processes involved in this context. In contemporary music therapy, the effective integration of technology is a significant concern. This paper aims to report on the Participatory Design process used to adapt the Emobook Life Story Book App for a Reminiscence Music Therapy Program for people living with dementia. An interdisciplinary team comprising an interaction designer, a software engineer, a music therapist, and a research assistant engaged in co-design meetings. The interdisciplinary and participatory process yielded four ideas and four lines of action, which evolved iteratively as the work meetings progressed. Points raised by the lines of action were addressed, leading to modifications in the Emobook app and adjustments in the intervention program for seamless integration. Collaborative, interdisciplinary efforts are essential in advancing the incorporation of technology into music therapy practice. This study demonstrates the value of a Participatory Design approach in adapting technology for use in Music Therapy.

Keywords: Music Therapy · Life Story Book · Participatory Design

A. L. Brooks (Ed.): ArtsIT 2023, LNICST 564, pp. 16–26, 2024.
https://doi.org/10.1007/978-3-031-55319-6_2

1   Introduction

In the field of contemporary music therapy, a notable concern refers to the effective incorporation of technological tools in professional work settings. The discipline has yet to reach a stage where it can be confidently stated that its practitioners have received comprehensive and adequate training for the competent use of technology within their practice [1]. In this context, music therapists are encouraged to network with experts from a wide range of disciplines, thus enhancing the development of their knowledge and skills. The synergies that emanate from such collaborative initiatives not only accelerate the assimilation of technological competencies but also generate knowledge that can be seamlessly integrated into music therapy practice. In other words, the path toward a professional practice that adequately integrates technologies is enhanced by collaborative and interdisciplinary work [1].

This paper aims to provide a comprehensive account of the Participatory Design (PD) process employed to adapt Emobook [2], a life storybook app designed for reminiscence therapy, for integration into a specialized Reminiscence Focused Music Therapy Program for people living with dementia [3]. To do this, an interdisciplinary team was formed, consisting of an interaction designer, a music therapist, a software engineer, and a research assistant. They conducted a series of co-design meetings that mixed meeting discussions, traditional material enactments, and mock-up screen design. These meetings served to elicit and validate requirements and produce a post-prototype of the life storybook app adaptation to be used in a music therapy program.

2   Related Work

2.1   Dementia Care

Dementia affects approximately 50 million individuals globally, with an annual increase of 10 million new cases. Projections suggest that by 2030, there will be 82 million people living with dementia, rising to 152 million by 2050 [4]. Given this growing prevalence, cost-effective and straightforward interventions are needed to enhance the quality of life for dementia patients and their caregivers.

Non-pharmacological approaches, such as reminiscence therapy, have demonstrated positive impacts on cognitive function, reduction of depressive symptoms, and promotion of positive self-esteem [5]. Particularly, music therapy has shown promise in improving mood, particularly depressive symptoms, in dementia patients in care settings [5, 6]. However, questions remain about the duration of these effects and the necessity for ongoing stimulation for lasting benefits [6].

2.2   Reminiscence and Music Therapy

Music therapy often triggers reminiscence naturally through familiar songs, facilitating conversations and the sharing of memories tied to personal and cultural identities. Research has demonstrated the positive effects of combining music and reminiscence therapy in reducing depressive symptoms among people living with dementia [7].

Music and reminiscence are often combined by a variety of practitioners in their work with people with dementia. Among these Music therapy techniques are:

- **Associative Mood and Memory Training (AMMT).** Cognitive rehabilitation technique that uses music to enhance memory processes in three ways - by producing a mood-congruent state to facilitate memory recall, by activating associative mood and memory networks to access long-term memories, and by instilling a positive mood at both encodings and recall enhancing learning and memory function [8]
- **Reminiscence Focused Music Therapy (RFMT).** Combined intervention with the use of associative items. The familiar music provided a supportive framework and acted as an anchor during periods of disorientation, directing the members of the group back to the present moment. The incorporation of associative items in the sessions encourages reality orientation, increased verbal interaction, and cognitive stimulation [7].

The two techniques offer a systematization of the reminiscence-focused process, including guidelines for the selection of music, session planning, and associated protocols [7, 8]. In particular, the AMMT involves a detailed clinical protocol and states that clinicians should discretely monitor patient behavior for affective response during music listening.

2.3 Reminiscence and Life Story Books

Life Story books are widely used to support reminiscence interventions [2]. These books typically contain photos or images of life memories. These help people living with dementia in constructing narrations about their life memories, which in turn has been shown to have positive effects on feelings of well-being and quality of life.

Emobook is a digital life story app designed to support reminiscence therapy. It aims to help therapists labor by giving them more control over multimedia stories and photos so that the reminiscing experience becomes stimulating and interactive (i.e., music, sounds, movie clips) and capturing activity records, which opens the opportunity for therapists to study the progress of the disease [2]. Its features include multilingual support, enable/disable settings for higher flexibility of the tool, full-screen support, and several mood meters to gather emotional responses from people living with dementia.

The utilization of such technologies in music therapy remains relatively underrepresented in the scientific literature. There is a notable scarcity of studies that specifically document the intricate design processes associated with these endeavors.

2.4 Participatory Design Methodology

Participatory design (PD) is a methodology that promotes the participation of users in the design process of health-related applications [3, 9]. It is an iterative process where each phase is planned by reflecting on the results from the previous phase concerning the participant's contribution.

Key activities of a PD process include fieldwork; literature reviewing; and development and testing. All activities must be applied with a participatory mindset that will ensure genuine participation throughout the project. For each of these activities, various methods are applied throughout the phases of a PD project to enhance user participation. These can be categorized under the headings of telling, making, and acting [3]:

- **Telling.** Practitioners are given a chance to share their knowledge.
- **Making.** Tools allow the ability to create, for example, through creative workshops conducted to generate ideas.
- **Acting.** Explore how new designs could affect practice, for example, by testing ideas in laboratories.

3 Interdisciplinary Participatory Design Process

The objective of this PD process was to merge different knowledge domains to co-design a post-first prototype of Emobook. It is worth highlighting that, given the application's purpose of aiding caregivers in the advancement of reminiscence therapy, the primary end users in this specific case were the music therapist and the research assistant. The process was developed in five working meetings. Each one was planned by reflecting on the results from the previous one. Key activities included fieldwork; literature reviewing; and development and testing. Below, we describe its attributes, encompassing the setting, team composition, tools employed, procedural steps, materials, schedule, and evaluation methods.

3.1 Setting

The first meetings were held online through the Teams platform, with access provided by the University of Santiago de Compostela. Starting with the second meeting, the meetings were held onsite at the CiTIUS (Singular Research Center for Intelligent Technologies), an institution belonging to the same University.

The meeting room was well-equipped to facilitate collaboration and productivity. It featured a spacious worktable, synchronized screens that mirrored the designer's tablet, and individual tablets for therapists' use. Additionally, traditional tools like paper and pens were readily available for sketching diagrams or notes. To ensure comprehensive documentation and continuity, we maintained diaries where we recorded key discussion points and identified areas for further exploration after each meeting. This blend of modern technology and traditional notetaking ensured a comprehensive and effective working environment.

3.2 Team

The team comprised four members: a music therapist responsible for designing and implementing the intervention program as part of her doctoral studies a research assistant with institutional experience and participant familiarity; the software developer and creator of the Emobook application; and a computer engineer and project collaborator. Additionally, the process received oversight from an experienced music therapist and the director of the Doctoral Program.

3.3 Tools

The co-design tools were thoughtfully selected to effectively amalgamate knowledge from diverse fields and foster collaborative problem-solving:

- **Brainstorming:** Within this tool, the "technology domain" encompasses emerging technologies that may stimulate innovative design concepts. Simultaneously, the "social health domain" focuses on health aspects open to potential adaptations.
- **Timeline:** These minute-by-minute diagrams outline the actions to be executed during the intervention. They prove invaluable in aligning the application's adjustments with the specific phases of the therapeutic process.
- **Work Diaries:** These serve as comprehensive records, documenting the subjects discussed, ideas generated, and future action plans emerging from meetings.

3.4 Procedure

Five meetings were convened at biweekly intervals, lasting approximately one and a half hours each. These work meetings followed a structured format divided into three distinct phases:

1. **Introduction (Telling).** The meeting began with an introduction to the topic at hand. This phase often involved the presentation of findings from literature reviews or research results, and it also included updates to the project timeline.
2. **Brainstorming and Idea Generation (Making).** The second phase involved a brainstorming meeting aimed at finding solutions to any issues or challenges that surfaced during the presentation. This creative process generated a pool of ideas and potential solutions.
3. **Action Plan (Acting).** The final phase focused on defining concrete steps and action plans for upcoming meeting.

Table 1 provides a concise overview of the meeting objectives. Meeting specifics are delineated in Sect. 4.

Table 1. Meetings objectives.

Meeting	Objective
1	Introduction of music therapy program with Emobook
2	Presentation of minute-by-minute timeline and assessment scales
3	Sharing results from literature review on emotional assessment scales
4	Integration of Emobook into intervention timeline
5	Discussion of configured Emobook post-prototype

3.5 Strategic Meeting Design and Evaluation

Each meeting was strategically designed to build upon the outcomes of the previous one. This deliberate sequencing aimed to facilitate the seamless integration of idea generation and the formulation of action plans, enabling well-informed decision-making for the successful Emobook integration within the music therapy program. To assess the effectiveness of our PD process, we considered several critical factors: the quantity and quality of ideas generated, their alignment with the defined lines of action, and their overall influence on shaping the new Emobook prototype.

4   Results

In this section, we focus mainly on presenting the outcomes of each meeting that followed the PD process. Moreover, we also present the process evaluation (Sect. 4.2).

4.1   Participatory Design Process

Meeting 1

Introduction. The music therapist outlined the program's objectives and components, designed under the AMMT technique, which includes the detailed clinical protocol [8]. Additionally, she clarified that the implementation would take place within a group setting, involving three older adults living with dementia, with a specific emphasis on social interaction outcomes [6]. The research assistant provided details about the institutional context.

Brainstorming. The technologists asked questions about the music therapist's presentation, the technique, and the processes involved. They also requested a copy of the protocol involved. They were interested in understanding the different stages of the session.

Action Plan. For the next working meeting, the music therapist should develop a minute-by-minute timeline of the intervention and state: what questions will be asked to the patients, when measures will be taken, and what instruments will be used (LA1).

Meeting 2

Introduction. The music therapist presented a minute-by-minute timeline in accordance with the clinical protocol, ensuring precise data correlation to meet the technical team's requirement for a comprehensive schedule. Subsequently, the technologists explained how to choose and preset the emotional assessment scales included in the app [2]. The data collection architecture of the application was elucidated, focusing on how Emobook facilitates the organization of memories, incorporating pictures or videos, into themes/chapters. This feature includes the ability to attach audio to each memory, intending to construct the life story of an individual user through these curated chapters.

Idea Generation. After listening to the clinical and technical explanations, the team came up with an idea to adapt Emobook, since the modality would be in a group of three participants. The first idea generated was: to group individual profiles into groups of three (IG1). Given the group setting, the chapters will encompass shared memories, featuring a combination of individual and group experiences.

Action Plan. Since the application allows selecting between different scales to assess emotions, it was considered necessary to carry out a literature review to find out which scales are most convenient for each moment and why. To undertake this endeavor, the music therapist and the assistant were tasked with a concise umbrella review, guided by the technical team members. This approach, distinct from a systematic review, was initiated based on the technical team members providing pertinent literature relevant to the subject. The line of action derived from this meeting was: to carry out a literature review of the emotional assessment scales included in the application to analyze which ones should be used and when (LA2).

Meeting 3

Introduction. The umbrella literature review uncovered various mood meters, ranging from simple like/dislike models to more intricate frameworks such as Plutchik's 8 basic emotions and Desmet's Pick-a-Mood to the Dementia Mood Picture Test [10–12]. It is worth noting that a preceding study that utilized Emobook supported the use of simplified mood meters for the precise capture of responses from individuals with dementia through direct questioning [2]. Moreover, therapists engaged in that study suggested incorporating a secondary mood meter to encompass a broader spectrum of emotions, thereby enhancing the emotional assessment process [2]. In response to this recommendation and the literature review, the music therapist proposed integrating a secondary mood meter for comprehensive emotion assessment.

Idea Generation. It was determined that the Ekman Scale would serve as the primary mood meter, employed each time a multimedia file was utilized to evoke a memory. This selection is based on its simplicity, as it offers fewer options compared to the other mood meters included in the app. Conversely, the Pick a Mood scale was suggested as the Secondary Mood Meter at the beginning of each session through the question "How are you feeling today?" to evaluate the mood before the intervention. The Pick a Mood User Manual advises that "given the relative stability of mood, it is advisable to inquire about it no more than twice around the event of interest" [10]. It also emphasizes that reporting moods immediately after an emotional event, such as a social interaction, may lead individuals to express their feelings about that anticipated event rather than their current mood [10]. Therefore, the idea generated in this meeting was to incorporate two emotional assessment scales into the music therapy program (IG2).

Action Plan. Consequently, for the next meeting, the action line involved creating a new timeline incorporating the use of Emobook, taking into consideration each of the moments of the meetings and the use of the assessment scales included in the app (LA3).

Meeting 4

Introduction. The music therapist presented the intervention timeline with Emobook integrated as a tool. In the second part of this meeting, the technologists showed how a smartwatch with a built-in scale would work to assess the emotional responses of people with dementia. Such a smartwatch interface suffers from touch overshooting and divided attention as the user has to look down the wrist for longer and maintain the arm in an inconvenient position. Further, unlike the tablet, the smartwatch would not allow the multimedia material to be manipulated, it would only incorporate the emotional assessment scale.

Idea Generation. Following a constructive exchange of viewpoints and a thorough review of the intervention timeline, a consensus emerged. It was collectively decided that equipping both the music therapist and the assistant with tablets featuring the new Emobook prototype was the optimal choice. The primary rationale behind this decision was to ensure that the assistant also had access to the application and an updated data backup, enhancing the overall effectiveness of the intervention. Therefore, the idea generated was related to the device selection: to use two tablets featuring the Emobook post-prototype for the music therapy pilot program (IG3).

Action Plan. One of the technologists commented that it would be convenient to have at least one additional measurement instrument or tool for more objectivity regarding the data collected. The music therapist recalled that the data could also be correlated with a music therapy assessment tool that would be used. The line of action derived from this proposal was to determine how the validated music therapy assessment tools complement the records obtained through Emobook (LA4).

Meeting 5

Introduction. Concerning the line of action that emerged in the previous meeting, the music therapist presented MiDAS [13], a validated music therapy tool developed to measure the observable musical engagement of persons with moderate or advanced dementia who may have limited verbal skills to directly communicate their musical experiences. "MiDAS focuses on capturing what people with dementia value in music (the "enjoyment")" [13] using a Visual Analogue Scale (VAS). It also allows qualitative data to be recorded.

Idea Generation. After the music therapist's presentation, it was decided that it would be positive to correlate the data obtained through the Emobook application with those obtained from the MiDAS assessment toolkit (IG4). Especially, considering that it allows qualitative data to be recorded, a relevant aspect facing the challenges and limitations of quantifying significant emotional experiences. In this sense, this assessment tool is adequately complemented by the scales incorporated in Emobook.

Action Plan. Finally, the technologists delivered the configured equipment with Emobook post-prototype. They explained that they configured the package to be updatable without deleting data, although previous tests must be carried out by creating three profiles. At the end of this meeting, the process was concluded with a post-prototype of Emobook ready to be tested by the music therapist before the start of the pilot program.

4.2 Process Evaluation

The results of the PD process include the generation of four ideas and four lines of action. The iterative condition of the process was reflected in the concatenation of LA and IG, as shown the Fig. 1.

As the work meetings progressed, several enhancements were identified for Emobook, including: (1) the development of a detailed schedule to ensure precise correlation of intervention-related data, (2) the addition of a secondary mood meter to enhance emotion registration, (3) the decision to correlate data with MiDAS Assessment Toolkit, (4) the configuration of Emobook Post-Prototype for group testing.

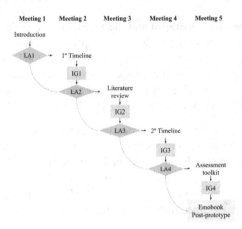

Fig. 1. PD process flowchart: lines of action (LA) and ideas generated (IG) achieved.

These enhancements collectively contribute to the integration of Emobook in a music therapy pilot program.

5 Conclusions and Future Work

The interdisciplinary collaboration in the participatory design (PD) process proved fruitful, leading to valuable adjustments in the Emobook post-prototype and its integration into reminiscence-focused music therapy programs. The collaborative process accelerated the assimilation of technological competencies, yielding knowledge seamlessly integrated into music therapy practice. Despite significant progress, the study emphasizes the need for further exploration, acknowledging the limited scope in connecting the experience with existing music therapy (MT) or technology literature. Challenges in literature research were acknowledged due to several unknown topics, influencing study decisions, highlighting the scarcity of scientific literature on the specific subjects in the initial review. Nevertheless, the study contributes valuable insights to the intersection of music therapy and technology.

The next phase involves initiating a pilot study to assess the viability of the new Emobook version, monitoring mood effects during and after the intervention, evaluating persistence, and determining the need for continued treatment. Additionally, assessing therapist acceptance, especially for this version in music therapy, is a key objective. The approved protocol for evaluating Emobook's efficacy in Music Therapy for older adults living with dementia has received Ethics Committee approval from the University of Santiago de Compostela (USC53/2023, December). The protocol ensures adherence to rigorous ethical standards. Integrating a qualitative research approach into the design and content analysis of life story books in reminiscence-focused music therapy settings will provide a deeper understanding. Advanced tools like NVIVO or ATLAS.ti will be employed for meticulous qualitative data analysis, enhancing the depth and reliability of findings.

Acknowledgements. This work was supported by the Ministry of Science, Innovation and Universities of Spain (grant PID2021-123152OB-C21) funded by MCIN/AEI/10.13039/5011000 11033, by "ESF Investing in your future" and the Galician Ministry of Culture, Education, Professional Training and University (grants ED431G2019/04 and ED431C2022/19 co-funded by the European Regional Development Fund, ERDF/FEDER program. Additionally, Noelia Gerbaudo has been a grant recipient for this paper by the José Otero-Carmela Martínez Foundation Grants for USC doctoral students 2023/24, carrying out her work in the Doctoral Program in Psychological Development, Learning, and Health at the Faculty of Psychology, University of Santiago de Compostela. We also extend our sincere gratitude to all the dedicated researchers and esteemed professors who played a pivotal role in shaping this work. We would like to express our special appreciation to Virginia Tosto, Music Therapist and Professor at the University of Buenos Aires, Argentina. Additionally, we are indebted to Romina Lizondo Valencia, who served as a Research Assistant during this project and was pursuing her master's degree at the University of Santiago de Compostela at that time.

References

1. Magee, W.: Music Technology in Therapeutic and Health Settings. Jessica Kingsley Publishers, London (2013)
2. Catala, A., Nazareth, D.S., Félix, P., Truong, K.P., Westerhof, G.J.: Emobook: a multimedia life story book app for reminiscence intervention. In: 22nd International Conference on Human-Computer Interaction with Mobile Devices and Services (MobileHCI '20). Association for Computing Machinery, New York (2021). https://doi.org/10.1145/3406324.341 0717
3. Clemensen, J., Rothmann, M.J., Smith, A.C., Caffery, L.J., Danbjorg, D.B.: Participatory design methods in telemedicine research. J. Telemed. Telecare **23**(9), 780–785 (2017). https://doi.org/10.1177/1357633X16686747
4. World Health Organization Homepage. Dementia Site page. https://www.who.int/news-room/fact-sheets/detail/dementia. Accessed 20 Aug 2023
5. Cammisuli, D.N., Danti, S., Bosinelli, F., Cipriani, G.: Non-pharmacological interventions for people with Alzheimer's disease: a critical review of the scientific literature from the last ten years. Eur. Geriatr. Med. **7**(1), 57–64 (2016). https://doi.org/10.1016/j.eurger.2016.01.002
6. van der Steen, J.T., Smaling H.J.A., van der Wouden, J.C., Bruinsma, M.S., Scholten, R.J.P.M., Vink, A.C.: Music-based therapeutic interventions for people with dementia. Cochrane Database System. Rev. **7** (2018). https://doi.org/10.1002/14651858.CD003477.pub3
7. Kelly, L., Ahessy, B.: Reminiscence-focused music therapy to promote positive mood and engagement and shared interaction for people living with dementia: an exploratory study. Voices: a World Forum Music Therapy **21**(2) (2021). https://doi.org/10.15845/voices.v21i2.3139
8. Thaut, M., Holmberg, V.: Handbook of Neurologic Music Therapy. Oxford University Press, Oxford (2014)
9. Vandekerckhove, P., de Mul M, B.W., de Bont, A.: Generative participatory design methodology to develop electronic health interventions: systematic literature review. J. Med. Internet Res. **22**(4) (2020). https://doi.org/10.2196/13780
10. Desmet, P.M.A., Vastenburg, M.H., Romero, N.: Pick-A-Mood Manual: Pictorial Self-Report Scale for Measuring Mood States. Delft University of Technology, Delft, NL (2016)
11. Clarke, C., et al.: Measuring the well-being of people with dementia: a conceptual scoping review. Health Qual. Life Outcomes **18**(1), 249 (2020). https://doi.org/10.1186/s12955-020-01440-x

12. Tappen, R.M., Barry, C.: Assessment of affect in advanced Alzheimer's disease: the dementia mood picture test. J. Gerontol. Nurs. **21**(3), 44–46 (1995). https://doi.org/10.3928/0098-9134-19950301-09

13. McDermott, O., Orrell, M., Ridder, H.M., et al.: The development of Music in Dementia Assessment Scales (MiDAS). Nord. J. Music. Ther. **24**(3), 232–251 (2015). https://doi.org/10.1080/08098131.2014.907333

Preliminary Findings from BehCreative: Exploring the Potential of Extended Digital Music Instruments for Music Therapy and Rehabilitation

Elena Partesotti[1(✉)], Gabriela Castellano[2,3], and Jônatas Manzolli[1]

[1] University of Campinas (UNICAMP), NICS, Campinas, Brazil
eparteso@unicamp.br
[2] University of Campinas (UNICAMP), GlebWataghin Institute of Physics, Campinas, Brazil
[3] University of Campinas (UNICAMP) Brazilian Institute of Neuroscience and Neurotechnology, Campinas, Brazil

Abstract. The usefulness of traditional musical instruments has already been demonstrated in music therapy and rehabilitation. In recent years, virtual reality systems have also been shown to promote good cognitive and motor rehabilitation results. Nevertheless, there are still few studies joining these things. The aim of this study was to demonstrate the application potential of an Extended Digital Musical Instrument (EDMI), BehCreative, in those areas, focusing on its ability to promote engagement, motivation, and confidence among users. For this, Gibson's concept of affordance to study users' creative behavior was used. Three healthy subjects participated in this study. Virtual Affordances (VAs) used by users during an exploratory phase and their Motion Development (jerk) were measured, and they answered the Affective Sliders self-assessment questionnaire. The results indicate a positive impact of BehCreative on emotional reactions, physical activities, and creative learning, opening avenues for future research and practical applications in the fields of motor learning and human-computer interaction in music therapy and rehabilitation.

Keywords: Extended DMI · Music Therapy · Creativity

1 Introduction

The use of Extended Digital Music Instruments in music therapy practice and rehabilitation is extremely limited or practically non-existent. In contrast, there exists a substantial body of literature on the application of music therapy in motor rehabilitation [1], including its use in Parkinson's [2] and post-stroke cases [3]. This can be attributed to the fact that music stimulates various brain reward systems, thereby enhancing the physical and cognitive performance of clients. Concurrently, the utilization of Virtual Reality (VR) in the realm of rehabilitation has witnessed a significant upsurge over the past decade [4].

Published by Springer Nature Switzerland AG 2024. All Rights Reserved
A. L. Brooks (Ed.): ArtsIT 2023, LNICST 564, pp. 27–40, 2024.
https://doi.org/10.1007/978-3-031-55319-6_3

Music therapy, recognized internationally in the 80s as a discipline capable of enhancing clients' physical and psychological well-being [5], centers around the active incorporation of traditional musical instruments to accompany and engage in dialogue with clients during sessions. This involvement may be led by one or more music therapists, depending on the applied methodology. In recent years, music therapy has increasingly intersected with the field of neurotechnology, involving the utilization of new technologies to ameliorate users' quality of life [6].

Despite this interdisciplinary impetus, research and implementation of DMIs in music therapy and physiotherapy have remained limited, with only a few examples to date [8, 9]. Considering the demonstrated effectiveness of music as therapy and the growing use of VR in therapeutic contexts, this paper aims to showcase the potential benefits that an Extended DMI can offer in these areas.

Within an EDMI the user becomes the instrument, as "the interaction occurs through a correlation between the environment and the agent, making the distributive nature of the space offered to the user essential as its property of manipulation and incorporation" [7].

The implementation of these technologies, in fact, could represent an excellent tool both for research and for caregivers during the delicate task of post-session analysis of clients' improvements.

This paper presents preliminary results of an exploratory study that tested the hypothesis that using an EDMI would increase the creativity and expression of the users, as measured by their motion development (Jerk) and Virtual Affordances.

2   Background

In this Exploratory Study we propose BehCreative as an EDMI [10]. In particular, within this EDMI, we considered the perspective offered by Embodied Cognition (EC) in designing a mapping closely connected to the movement and gestures of the subject in the environment, in which the body itself becomes a musical instrument. Within the EC paradigm, in fact, the common denominator is the role played by the body in experiencing and perceiving both the inner and the outer world in an enactive continuum [11]. EC contains various theories within it, many of which have been studied in recent years and are developing further. When dealing with an EDMI, we should also keep in mind Gibson's theory of visual perception, in particular the concept of Affordance that refers to the characteristics of the environment offered to the subject [12]. Affordances represent the potential of an object to be used in a particular way and for a particular purpose, a kind of invitation to use an object in a specific way.

In BehCreative (Fig. 1) there are no physical objects, but rather virtual possibilities depending on audiovisual feedback decided *a priori* through mapping. From the point of view of mapping, this virtual gesture refers to the concept of Multiple Affordance (MA) and Sensorimotor Map (SM) - somewhere else described [9] - and are determined by the subject's exploration of the environment. While MAs are the possible paths made available to the subject depending on the interaction with the EDMI and defined *a priori* by the mapping, the user can change the type of interaction or the ways they are used, following a specific goal within the environment by creating always new SMs. And this characteristic depends on the mapping of the EDMI. At the same time, this goal is linked to the Creative Empowerment that "happens when the subject gains full control over the technology and is therefore capable of creative expression, self-control, and awareness" [10]. Hence, giving these considerations, we believe that BehCreative is a potentially promising EDMI with creative and expressive possibilities, grounded on the explorative behavior of the users and Creative Empowerment.

This research was interrupted due to the COVID-19 pandemic; therefore, of the 10 volunteers who took part in the experiment, only 3 completed all sessions and their data are presented in this article.

Fig. 1. BehCreative EDMI usage during an experiment with two users.

3 Material and Methods

We conducted an exploratory study with 3 healthy volunteers, between 21 and 38 years old, 1 male. The project was approved by the Ethics Committee of University of Campinas and all subjects gave their written informed consent before entering the study. For the experiment, the subjects explored the BehCreative environment for 10 days (during two weeks, from Monday to Friday) for a maximum of 5 min, without receiving any particular instruction, on the contrary, having maximum freedom to interrupt the experiment if they wanted. Each subject had a different musical and physical background; they were selected having in mind a future formation of the following groups:

- Control Group - subject with no previous knowledge of music and a regular fitness style of life;
- Music Group - subject with basic music knowledge;
- Exercise Group - subject with dance or sport background, but preferably with no musical background.

Before and after each session, the subjects completed an evaluation using Affective Sliders. The arousal and pleasure values were derived using the Affective Sliders method developed by Betella and Verschure [13]. This method involves utilizing a set of sliders to quantify the emotional dimensions of arousal and pleasure experienced by subjects during interactions with a given stimulus or experience. Participants provide real-time feedback by adjusting the sliders along a continuum, reflecting their emotional states. The Affective Sliders provide a structured and quantifiable way to assess emotional responses, enabling the measurement of subjective emotional experiences in a controlled experimental setting.

3.1 Equipment

BehCreative (Behave Creatively) is an EDMI developed within the NICS laboratory (Interdisciplinary Nucleus for Sound Studies) and BRAINN (Brazilian Institute for Neuroscience and Neurotechnology), at University of Campinas. In the first version of this EDMI proposed here, BehCreative consisted of an immersive environment inside the NICS recording studio made up of an octophonic system for sound diffusion, and three screens in which visual feedback connected to the sound is projected (in front and to the sides of the user). For mapping, Kinect 2 made it possible to use several programs - Processing, Pure Data, Sadie - connected via OSC (Fig. 2).

The audiovisual feedback consisted of different types of sounds and visuals that changed according to the jerk of the subject. Jerk is the rate of change of acceleration and it is relevant for measuring the movement of the subjects because it reflects their speed, intensity, and direction. The sounds were synthesized using Pure Data and they varied in pitch, timbre, duration, and harmony. The visuals were generated using Processing and they varied in shape, size, color, and brightness. The mapping between the gestures, sounds, and visuals was based on the concept of MA and SM quoted above.

We implemented upper-limb gestures as MAs - that is to say, a specific gesture as a virtual command in order to play the instrument that is the user's body itself - and these were named Virtual Affordances (VAs) (Fig. 3). Thus, in the mapping, we added six VAs [8] - feedback connected to pre-established gestures of the subjects, which corresponded to certain sounds lasting a few seconds feedbacks and predetermined colors. At the same time more jerk corresponded to dissonance and black and white. Figure 3 shows an example of how the mapping worked for one VA.

3.2 Data Collection and Analyses

We collected data from three sources: Kinect 2, Pure Data, and a self-assessment questionnaire. Kinect 2 recorded the position and movement of the subjects' joints in three-dimensional space, with an acquisition rate of 1 Hz. Pure Data recorded the sounds

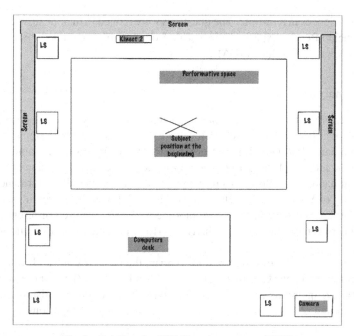

Fig. 2. BehCreative EDMI environment at NICS.

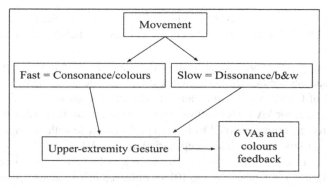

Fig. 3. Visual Mapping with correspondence of Movement. Fast movements correspond to consonant sounds and coloured particles screens. Slow movements correspond to dissonant sounds and black & white particles. Moreover, six VAs corresponds to six upper-limb gestures.

produced by the subjects and their corresponding jerk levels, also with an acquisition rate of 1 Hz. The sessions had a maximum duration of 5 min.

The self-assessment questionnaire consisted of two questions that asked the subjects to rate their experience with BehCreative using the Affective Slider scale from 1 to 5. The Affective Slider, developed by Betella and Verschure [13], is a tool to assess affective responses in participants during an interaction with a particular stimulus or experience. It is a continuous rating scale that allows individuals to express their emotional states or affective experiences in real-time. The questions were related to their positive interaction self-assessment after the experience in a scale from 1 to 10. We named this measure as Experience Appreciation Index (EAI).

3.3 Data Analysis

We used descriptive statistics to summarize the data from each source. We calculated the mean and standard deviation of the jerk levels for each subject, each session and for both left and right hand. For the Virtual Affordance (VA), we employed the same approach, calculating the mean and standard deviation for each subject, each session and for both left and right hand. This allowed us to gain insight into the overall performance trends and variability of VA and Jerk levels within each subject over the 10 sessions. Next, we used the Matplotlib library in Python to plot the bar chart. We used different colors to distinguish between the subjects and the two performance measures (VA and Jerk), see Fig. 3.

In this way, we aimed to visualize and compare the mean VA and Jerk values across three types of participants: Control (C), Music (M), and Exercise (E) over the course of 10 sessions. This visual representation complemented the descriptive statistics, enhancing our understanding of the participants' performance across the 10 sessions.

4 Results

The graphs in Fig. 4 show the average values of VA and Jerk for the three participants: CS, MS, and ES. The data is divided into 10 sessions numbered from 1 to 10 on the x-axis.

Throughout the sessions, we can see how the values of VA and Jerk change among the different subjects, indicating performance differences between them.

The trend of CS shows opposite values in the first six sessions of VA and Jerk: a decrease (increase) for VA (Jerk), followed by an increase (decrease) and then another decrease (increase). From session 7 to 10, two peaks alternate with two drops, for both VA and Jerk, but Jerk displays less abrupt variations than VA.

MS exhibits a similar trend between VA and Jerk in the first four sessions, followed by some regularity in Jerk (sessions 4 to 10) not observed in VA, which instead, displays some extreme variations.

Finally, ES demonstrates consistent Jerk and VA patterns of values, except for larger VA values in sessions 5 to 7, nonetheless making ES the most consistent subject.

Regarding VA, the peaks may indicate moments of intense exploration, while lower values may suggest usage driven by learning between gestures and audiovisual feedback. In this perspective, CS continues exploration even in the last sessions, while MS and ES exhibit more stable behavioral learning.

Considering graphs in Figs. 5 and 6:

Mean VA for each subject and session

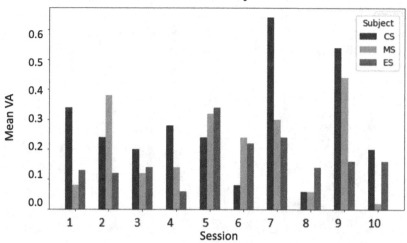

Mean Jerk for each subject and session

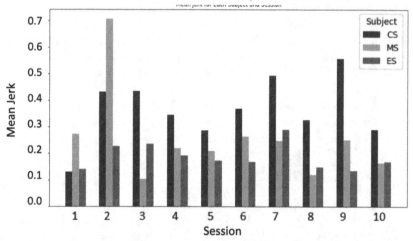

Fig. 4. Average values of VA (upper graph) and Jerk (lower graph) for the three participants: C, M, and E. The x-axis of the charts represents the 10 sessions, numbered from 1 to 10. The y-axis shows the range of mean values for each performance measures.

Arousal Trends

Prior to the experiment, participants exhibit varying levels of arousal. The control subject (C) shows a fairly wide range of arousal, with some sessions indicating higher excitement and others less. The musician subject (M) generally maintains elevated levels of arousal, suggesting emotional engagement and heightened attention. The exercise subject (E) initially displays high arousal levels, which could be associated with the anticipation of physical exercise, and then decreases in the later sessions.

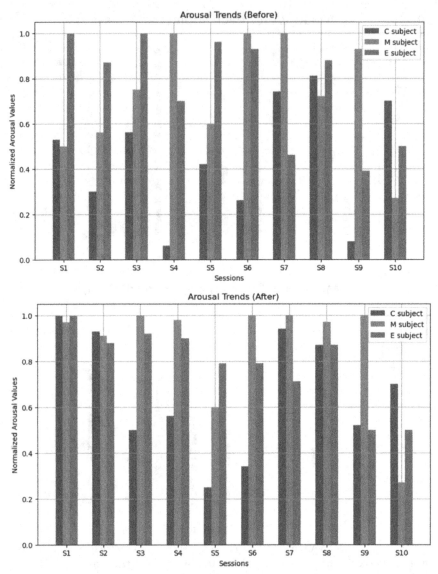

Fig. 5. Values of Arousal before and after the experiment, following the Affective Slider self-assessment.

Following the experiment, subjects appear to have regulated their arousal levels differently. The control subject shows a more consistent trend compared to the beginning, with higher excitement in 7 out of 10 sessions, possibly indicating increased stimulation. The musician subject (M) maintains elevated levels of arousal even after the experiment, implying sustained engagement. The exercise subject (E) seems to have balanced arousal regulation, likely in response to movement.

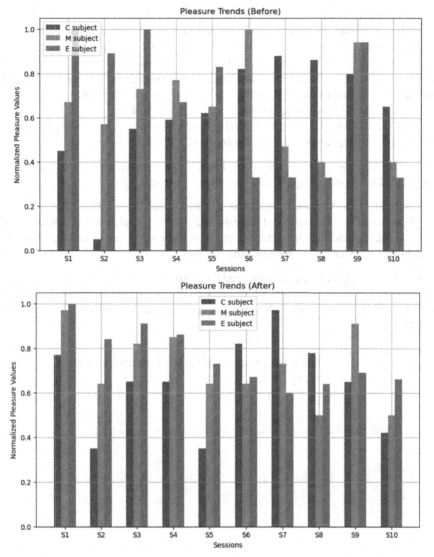

Fig. 6. Values of Pleasure before and after the experiment, following Affective Slider self-assessment.

Pleasure Trends

Before the experiment, pleasure levels vary among subjects. The control subject shows increasing pleasure values over sessions, while the musician subject and the exercise subject exhibit high values at the first sessions with an abrupt drop in the last sessions (with exception of session 9). This might reflect a positive expectation towards the experiment.

Post-experiment, pleasure levels tend to stabilize for the musician (M) and exercise (E) subjects. However, the control subject's pleasure increases only in half of the sessions, possibly linked to a more positive experience with EDMI usage. The musician (M) and exercise (E) subjects maintain a high level of pleasure, increasing the values in the last sessions, suggesting a rewarding experience. For E, this may indicate that the sensorimotor learning experience through movement has been positive.

The EAI (Fig. 7) provides subjective feedback from participants about their engagement with BehCreative. Looking at the EAI questionnaire results, we can observe certain trends and differences among the three subjects (C, M and E):

Control Subject: Participant C has scores for half of the sessions with low values (less than or equal to 3), indicating the most unstable response.

Music Subject: User M showed variations in ratings across the sessions. The user's positive self-assessment process showed some fluctuations. This could suggest that users with a musical background may have a more nuanced response to the technology, possibly due to higher expectations or specific preferences.

Exercise Subject: User E consistently rated their experience with BehCreative very positively, giving high scores throughout the sessions. The user's responses were relatively stable throughout the sessions, indicating a consistent level of positive experience with the EDMI.

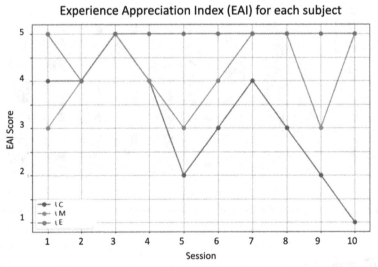

Fig. 7. The graph illustrates the participants' subjective feedback. The x-axis represents the different sessions (from S1 to S10) in which the participants interacted with BehCreative. The y-axis represents the Experience Appreciation Index (EAI) scores provided by the participants.

5   Discussion

From the analysis of the graphs, we can make some hypotheses about the effect of different experimental conditions on the variables VA and Jerk. Overall, the analysis of Mean Jerk values reveals varying trends in motor control and movement smoothness across the three subjects. The CS subject exhibits more irregular Jerk patterns, potentially reflecting challenges in motor coordination. On the other hand, the ES subject demonstrates more consistent and controlled movements, implying better motor performance. The MS subject falls in between, with sessions of both smoother and less smooth movements. These differences may indicate variations in motor learning and sensorimotor adaptation among the subjects.

The observed differences between subjects and the variations in VA and Jerk can be attributed to the participants' different movement experiences and needs, given their different backgrounds. C shows evident fluctuations in Mean VA values but less so for body acceleration (Jerk). M also shows fluctuations in Mean VA values but with smaller amplitudes, and an initial trend of growth in Mean Jerk values. On the other hand, E shows a steady increase in VA values and a (low amplitude) large stability in Mean Jerk values.

Correlating the results of pleasure with those of arousal, several interesting observations can be made. For instance, it can be noted that the increase in pleasure for participants M and E after the session (Fig. 6, bottom) appears to align with the rise in arousal (Fig. 5, bottom). This suggests a potential relationship between pleasurable experience and heightened arousal for M and E subjects.

From the results, indeed, it appears that emotional engagement and subjective experience during EDMI usage are influenced both by initial arousal levels and changes in pleasure during the experiment. Subjects responded differently to sensorimotor learning sessions, with the musician subject (M) displaying consistent engagement and the exercise subject (E) adjusting emotional levels in response to physical activities. The EDMI seems to have a positive impact on participants' emotional experience, contributing to their interaction and engagement in learning to use the body as a musical instrument.

The consistently positive ratings of EAI for the M and E subjects (and for half the sessions for the C subject) indicate that the EDMI has the potential to enhance users' engagement, expression and thus satisfaction. The user with a physical exercise background (E) appeared to have particularly positive experiences, which aligns with the paper's focus on therapeutic and music therapy applications.

The questionnaire (EAI) results also support the paper's discussion on the impact of musical experience and physical training on users' responses to the EDMI. Additionally, the variations in ratings suggest that the technology-based approach may cater differently to individuals with different backgrounds, emphasizing the need for personalized approaches in therapeutic settings. In conclusion, the data obtained from the experiment seem to suggest that the use of BehCreative has an impact on the emotional reactions and physical activities of the participants. The results indicate that there may be significant differences in responses among people with the three types of backgrounds analyzed during the interaction with the immersive musical experience. These differences may be relevant for the creative learning and musical expression of the subjects.

The results from this study contribute - despite the low statistical power of the data - to the growing body of research on motor learning and human-computer interaction, providing valuable implications for future studies and practical applications in related fields.

5.1 Future Perspectives

The interpretation of the results is an important aspect to understand the impact of EDMIs on subjects' behavior and for future directions. It could be interesting to analyze the differences in movement types and how they are linked to the internal MS (see Sect. 2). Furthermore, the Mean Jerk indicates the overall body acceleration during the experiment and could be correlated with the complexity of movements or the expressiveness of the subjects. Moreover, the audio and visual feedback provided by the EDMI may have had a significant impact on the subjects' behavior. For instance, the visual feedback projected on three screens might have influenced their perception of space and interaction with the surrounding environment. Regarding the impact of training and experience, the difference between C's and M's responses could be related to the musical training of subjects in the latter case. It could be interesting to examine how musical experience influences the interpretation and use of the EDMI in future research. Finally, regarding creative potential and learning, BehCreative appears to offer new opportunities for expressing Creative Empowerment and learning the EDMI. The ability to manipulate sound and visual feedback could stimulate artistic exploration and the learning of new musical techniques.

The study's limitations include the small sample size and the interruption due to the COVID-19 pandemic. Despite these limitations, results provide valuable insights that can inform future research with larger and more diverse populations.

To further understand the impact of the EDMI, future studies could explore correlations between EAI, Affective Sliders scores and users' creative output, motor learning progress, and emotional states. Additionally, incorporating qualitative interviews or focus groups could offer deeper insights into users' experiences and suggestions for improvement.

6   Conclusions

The results of the pilot study with the EDMI suggest that this technology can significantly influence the behavior of subjects belonging to different groups. The observed differences between the C, M, and E may indicate the impact of musical experience, physical training, and artistic skills on the response to the EDMI. The interaction between body movements, audiovisual feedback, and general motion development offers new perspectives for creative expression and musical learning. Furthermore, the Affective Sliders results on arousal and pleasure and the line plot of EAI scores connect the concept of Creative Empowerment and engagement with users' experiences while using BehCreative. The improved level of pleasure's responses after the experiment and the positive EAI scores indicate that the EDMI has the potential to empower users creatively and create an engaging musical experience, supporting its application in music therapy and rehabilitation contexts. The variations in arousal and pleasure among users, furthermore, underscore the importance of personalized approaches to optimize the therapeutic benefits of the EDMI for different individuals.

However, with only three subjects, it is important to remember that the results are not representative of the general population. Further studies with a larger sample are currently being conducted and analyzed to confirm or delve deeper into the observations obtained from this experimental study.

Acknowledgments. This research was supported by São Paulo Research Foundation – Brazil (FAPESP) – Grants 2016/22619-0 and 2013/07559-3 and Brazilian National Council for Scientific and Technological Development (CNPq) – Grant 308695/2022-4.

References

1. Weller, C.M., Baker, F.A.: The role of music therapy in physical rehabilitation: a systematic literature review. Nord. J. Music. Ther. **20**(1), 43–61 (2011)
2. Pacchetti, C., Mancini, F., Aglieri, R., Fundaro, C., Martignoni, E., Nappi, G.: Active music therapy in Parkinson's disease: an integrative method for motor and emotional rehabilitation. Psychos. Med. **62**(3), 386–393 (2000)
3. Jeong, S., Kim, M.T.: Effects of a theory-driven music and movement program for stroke survivors in a community setting. Appl. Nurs. Res. **20**(3), 125–131 (2007)
4. Feitosa, J.A., Fernandes, C.A., Casseb, R.F., Castellano, G.: Effects of virtual reality-based motor rehabilitation: a systematic review of fMRI studies. J. Neural Eng. **19**(1), 011002 (2022)
5. Nordoff, P., Robbins, C.: Creative Music Therapy: A Guide to Fostering Clinical Musicianship. Barcelona Pub., Barcelona (2007)
6. Brandão, A.F., et al.: Biomechanics sensor node for virtual reality: a wearable device applied to gait recovery for neurofunctional rehabilitation. In: Gervasi, O., et al. (eds.) ICCSA 2020. LNCS, vol. 12255, pp. 757–770. Springer, Cham (2020). https://doi.org/10.1007/978-3-030-58820-5_54
7. Partesotti, E.: Extended digital music instruments to empower well-being through creativity. In: Brooks, A.L. (ed.) Creating Digitally. Shifting Boundaries: Arts and Technologies - Contemporary Applications and Concepts, vol. 241, pp. 365–401. Springer, Cham (2023). https://doi.org/10.1007/978-3-031-31360-8_13
8. Partesotti, E., Peñalba, A., Manzolli, J.: Digital instruments and their uses in music therapy. Nord. J. Music. Ther. **27**(5), 399–418 (2018)
9. Partesotti, E., et al.: Analysis of affective behavior in the artistic installation MovieScape. In: Brooks, A.L. (ed.) ArtsIT. LNICSSITE, vol. 479, pp. 327–345. Springer, Cham (2023). https://doi.org/10.1007/978-3-031-28993-4_23
10. Partesotti, E., Feitosa, J.A., Manzolli, J., Castellano, G.: A pilot study evaluating brain functional changes associated to the BehCreative environments. J. Epilepsy Clin. Neurophysiol. **23**(3), 12 (2020). ISSN 16176-2649. In: Proceedings of 7th Brainn Congress, Brazilian Institute of Neuroscience and Neurotechnology
11. Peñalba, A., Valles, M.J., Partesotti, E., Sevillano, M.Á., Castañón, R.: Accessibility and participation in the use of an inclusive musical instrument: the case of MotionComposer. J. Music Technol. Educ. **12**(1), 79–94 (2019)

12. Porayska-Pomsta, K., et al.: Developing technology for autism: an interdisciplinary approach. Pers. Ubiquit. Comput. **16**(2), 117–127 (2012)
13. Betella, A., Verschure, P.F.: The affective slider: a digital self-assessment scale for the measurement of human emotions. PLoS ONE **11**(2), e0148037 (2016). https://doi.org/10.1371/journal.pone.0148037

A Systematic Review of the Technology Available for Data Collection and Assessment in Music Therapy

Beatriz Amorós-Sánchez[1]([✉]) [iD], David J. Gamella-González[1] [iD],
Pablo Cisneros-Álvarez[1] [iD], and Vicenta Gisbert-Caudeli[2] [iD]

[1] Universidad Internacional de la Rioja, Logroño, Spain
beatriz.amoros@unir.net
[2] Universidad Autónoma de Madrid, Madrid, Spain

Abstract. The lack of standardized tools for assessment in music therapy is a reality that has concerned music therapists for years. Considering technology as one of the advances that have most modified our lives in recent years, this study carries out a systematic review to analyze the technology available for data collection and assessment in music therapy. The databases analyzed were PubMed and Scopus, from 2013 to June 2023. Out of 370 records, only 8 records met the inclusion criteria, showing a strong prevalence in publications in 2022. This review suggests that the technology is increasingly present in music therapy sessions and it is also growing the number of studies worried to find more accurate solutions for data collection and assessment in music therapy. Taking advantage of technological resources for this process seems to be a great opportunity, but the research in this context is insufficient for this moment, so there remains too much work to do.

Keywords: Music therapy · Technology · Datta collection · Assessment

1 Introduction

In the past decades much music therapy research has focused on clinical practice and the experiences of therapeutic projects based on music therapy. Led by some of the best-known names in music therapy development, many of these investigations have focused on clarifying the benefits obtained from clinical work in cases of treatment of autistic spectrum disorder [1], for acquired brain injury [2], to slow the effects of dementia [3], among many others; either individually or for groups [4]. These studies are of vital importance to support the use of music as therapy, to continue growing in the discipline and to inspire other potential music therapy projects. For this aim, it is also important to ensure accurate tools that reflect the results of these practices [5] and this is an issue that has been worrying many of the music therapy studies, especially in the last years. Nevertheless, an evident lack has been identified in the music therapy research on music therapy data collection and assessment.

A. L. Brooks (Ed.): ArtsIT 2023, LNICST 564, pp. 41–54, 2024.
https://doi.org/10.1007/978-3-031-55319-6_4

In their proposal of a taxonomy of the qualities of music therapy interactions in cancer care, Magee, Davidson, and Hwang [6] especially emphasize the importance of accurately defining and measuring the qualities of music therapy interactions to better understand the outcomes of music therapy. Mössler, Chen, Heldal, and Gold [7] stress the rigorous use of precise measurement tools to assess outcomes after conducting a systematic review examining the efficacy of music therapy in people with schizophrenia and schizophrenia-like disorders. And, in the same way [8], insist during their meta-analysis on the importance of using precise research tools, such as activation probability estimation analysis, to better understand the neurological processes involved in musical improvisation in therapeutic contexts. Magree [9] further adds how data collection and assessment tools should reflect the artistic and unique nature of music therapy, being essential to find the accurate method.

We certainly need to be concerned about the importance of evidence-based practice in music therapy and how data collection and assessment tools need to be updated to reflect these advances in research and clinical practice [10], so we should aim to integrate contemporary perspectives in this process [11].

The most used tools are mainly inspired by the proposals of the first referents of music therapy, which established the working bases in the second half of the 21st century and which, therefore, have become obsolete with the passage of time, especially because they are not taking advantage of all the opportunities that technology offers us. It is true that data collection tools and assessment in music therapy have evolved in the past decades, trying to adapt to the advances in music therapy as analyzed by Schmid and Elefant [12], but observation is still used as the main method, and data collection is usually done with a paper and a pen [13].

We must develop standardized assessment tools to specific populations or needs to guarantee the quality and comparability of music therapy studies [14]. It is not essential to have the same assessment, because not all practices are the same, but it is essential to have standardized tools [15]. Another important point is to standardize the terminology used [16] and establish a common vocabulary to avoid mismatches in the interpretation of data.

In his systematic review, Gattino [13] registered a strong tendency to study assessment practices in children, possibly justified by the large number of music therapists who attend or investigate within this population or context, but he stated that there are no specific studies of how to use technology in a practical way in the context of the data collection and assessment in music therapy. And this article aims to shed some light in this regard.

2 Literature Review

The assessment in music therapy is an essential process, both to ensure the quality of work and to achieve a qualitative increase in the recognition of the discipline [17]. It is considered that data collection is the initial step of gathering information and assessment is a broader process that may involve data collection and evaluation but also encompasses making informed judgments and decisions in context. On the one hand, the assessment in music therapy allows the music therapist to know in detail the needs, abilities and goals of the people with whom they work, giving them some essential keys to be able to design an appropriate therapeutic process. On the other hand, it allows the results obtained with clinical practice to be endorsed, for which it is essential that there be rigor and adequate and objective evaluation bases. Whenever something is evaluated, it is necessary to use an appropriate evaluation method and, especially, one that allows us to measure what we want to measure. Sotelo [18] reflects that we cannot measure the temperature with a tape measure or the distance with a thermometer, although both are valid tools and the concepts are measurable. That is why it is important to choose well what you work with. The main challenge is to translate into words or data what happens in the sessions with music, especially with objectivity and rigor [19].

Despite this, university training programs often focus on practice and the fundamentals of practice and do not devote the necessary time to research, data collection and assessment. Almost 10 years ago, Benenzon [20] identified this lack in the training of music therapy students, highlighting a greater need to integrate research and assessment skills into university training programs; and it seems that reality has not changed as much as expected, following the criticism of Magee and Byrland [21] where they emphasize this same idea and the need for more comprehensive training that includes data collection and evaluation to support the effectiveness of music therapy, especially in medical contexts where these methodologies are frequently used.

It's very important to lay the knowledge foundations of music therapy data collection and assessment, to determine what is to be measured and how, but then it is essential that there be flexibility, because not all therapeutic processes are the same [22]. Since the use of music therapy was standardized, there have been a variety of evaluation methods and tools that seek to measure in the most reliable way the items on which work is done in the sessions of the discipline [23].

For this purpose, music therapy has frequently used some of the foundations and tools used for evaluation in other disciplines, combining some of these tools with its main method of data collection: the phenomenological observation. We know from Wigram and Wosch [24] how the phenomenological observation can provide valuable insight into the therapeutic process and in multiple clinical contexts. In their book *Microanalysis in Music Therapy: Methods, Techniques and Applications for Clinicians, Researchers, Educators and Students* they offer a "wide range of instruments, analysis tools, and aids for systematically supporting these perceptions through a multitude of microanalysis methods in music therapy for different problems in clinical practice, different client populations, and different approaches of music therapy". Within the 20 methods of microanalysis in music therapy thoroughly analyzed in this book, we see that there are different tools that stand out for data collection:

- Video recording: analyzing the video recording and later transforming it into a music notation and comments in the score [25], with a coding system that leads to graphic representation [26], in an excel sheet with established descriptions [27], standard items for behavior analysis [28] or in responses for a form analysis based on events [29], among others.
- Audio recording: analyzing the audio recording and building with it a graphical representation of the waveform [30], a two-dimensional spatial diagram generated with PrinGrid [31], among others.

- Written descriptions: text descriptions of the musical experience [32] or narratives about the experience and its reflections [33], among others.
- Scales: transformation of the sessions into qualitative scales with different levels of compliance with the analyzed behaviors [34], Likert scales or scales of musical parameters [35], among others.

Although most of the cases previously exposed focus on individual clinical work, in the case of community music therapy, Stige, Ansdell, Elefant and Pavlicevic [36] also explore phenomenological observation as a valid tool to understand the musical experience in this context.

Parallel to the establishment of these tools for data collection and microanalysis, technology has evolved immensely, offering new possibilities for music therapy. It has been 13 years since Ruud announced a new direction for music therapy, reflecting on how the everyday musicking through smartphones and mp3-technology may initiate a new area in the use of music as self-caring technology [37]. Furthermore, as indicated by Andrew Knight [38]: "since the release of the first iPad in 2010, development of software applications (apps) with uses for music therapists has increased dramatically. Apps for musical instruments, notation/composition, and recording/playback range from easy-to-use interfaces to highly technical apps for industry professionals". Ten years have passed since this affirmation and now some music therapists have also seen the possibility of taking advantage of this technology to improve assessment in music therapy, recently creating the following tools [23]:

- IMTAP (Individualized Music Therapy Assessment Profile) is am assessment protocol developed by a team of six experienced music therapists [39]. Now it is digitized and allows an automated process in music therapy. This tool, available in Portuguese now, has the following sections: admission form, planning and implementation, initial assessment, completion/discharge and data collection. The admission form allows to collect general information about the patient and information about sensory, communicative, cognitive, perceptive, emotional, and social skills. The initial evaluation is divided into an initial evaluation sheet, potentialities and challenges, and goals and objectives. And the data collection section is divided into gross motor skills, fine motor skills, oral motor skills, sensory, receptive communication, expressive communication, cognitive, emotional, social and musicality; and each section is valued with a scoring scale.
- MITM (Meet In The Music) by Esther Thane (n.d.), is a music therapy curriculum for autism and diverse needs that offers enormous resources for the music therapist: a music therapy goal bank, examples of interventions - with songs, audio files, picture cards, sheet music, training videos of music therapy strategies, parent programs to do at home...-, a method of qualitative and quantitative data collection for analysis, templates for treatment plans, among others. MITM even proposes to the music therapist activities and objectives, personalized for each patient by adding information about this person. Currently this tool is only available in English, but it is expected that a Spanish version will also be available soon and the author herself will provide training for music therapists to learn how to use it.

On the other hand, the technology applied both to music production and to the recording of patient responses plays an increasingly important role in the clinical context of music therapy. There are digital applications that support the sessions by generating music, making it possible for users who have certain physical or cognitive difficulties to participate. Among the multiple tools identified in this sense, two stand out that have been considered the most useful for the authors of this article:

- Virtual Musical Instrument (VMI): this digital program "allows the slightest of movements to produce music on a computer - bringing the empowerment of making music to those for whom picking up a guitar or sitting at a piano are activities inaccessible due to disability" [40]. It has been designed for therapists and educators and is customizable according to the preferences and needs of the user. It has been invented in Professor Tom Chau's Pediatric Rehabilitation Intelligent Systems Multidisciplinary (PRISM) lab at Holland Bloorview Kids Rehabilitation Hospital.
- SoundBeam: this tool is "a 'touch-free' musical device that turns physical movements into sounds and music. Using cutting-edge sensor technology, Soundbeam translates any physical gestures into musical notes, enabling people of all abilities to create music" [41]. This digital musical instrument also allows combining various sensors and configuring the sound sets and musical parameters to adapt them to the person's needs.

All these previously mentioned resources are considered of great value for the music therapist, but a greater knowledge about the technology available for data collection and assessment in music therapy could be of even greater value for the discipline. That is what this systematic review promotes.

3 Methodology

This report aimed to clarify the current technology available for data collection and assessment in music therapy. The research question is: how could take advantage of the use of advanced technology for data collection and assessment in terms of assessing the effectiveness, efficiency, and accuracy of practice in music therapists' sessions? To identify all potential tools, we screened the literature with a systematic review in PubMed and Scopus. A first systematic literature search was conducted on 30/06/2023 in the two databases selected, the systematic search was limited to systematic reviews in English or Spanish, published from 2013 to 2023.

We initially used the terms "music therapy" AND "technology" OR "virtual" OR "digital" - in English or Spanish - and we attained 821 results. Then we decided to narrow the search further and we added the terms "data collection" and "assessment". The inclusion criteria for relevant studies are summarized in Table 1.

Table 1. Inclusion criteria.

Heading level	PubMed and Scopus
Languages	English and Spanish
Date	30/06/2023
Publication period	2013 – June 2023
Search	**(music therapy OR musicoterapia) AND ((technology) OR (technological) OR (virtual) OR (digital) OR (tecnología) OR (tecnológico)) AND ((data collection) OR (assessment) OR (toma de datos) OR (registro de datos) OR (recopilación de datos) OR (evaluación))** ("music therapy"[MeSH Terms] OR ("music" [All Fields] AND "therapy" [All Fields]) OR "music therapy"[All Fields] OR "musicoterapia" [All Fields]) AND ("technology" [MeSH Terms] OR "technology" [All Fields] OR "technologies" [All Fields] OR "technology s" [All Fields] OR ("technologic" [All Fields] OR "technological" [All Fields] OR "technologically" [All Fields]) OR ("virtual" [All Fields] OR "virtuality" [All Fields] OR "virtualization" [All Fields] OR "virtualized" [All Fields] OR "virtualizing" [All Fields] OR "virtuals" [All Fields]) OR ("digital" [All Fields] OR "digitalisation" [All Fields] OR "digitalised" [All Fields] OR "digitalization" [All Fields] OR "digitalize" [All Fields] OR "digitalized" [All Fields] OR "digitalizer" [All Fields] OR "digitalizing" [All Fields] OR "digitally" [All Fields] OR "digitals" [All Fields] OR "digitization" [All Fields] OR "digitizations" [All Fields] OR "digitize" [All Fields] OR "digitized" [All Fields] OR "digitizer" [All Fields] OR "digitizers" [All Fields] OR "digitizes" [All Fields] OR "digitizing" [All Fields]) OR ("tecnología" [All Fields] OR "tecnologías" [All Fields]) OR "tecnológico" [All Fields]) AND ("data collection" [MeSH Terms] OR ("data" [All Fields] AND "collection" [All Fields]) OR "data collection" [All Fields] OR ("assess" [All Fields] OR "assessed" [All Fields] OR "assessment" [All Fields] OR "assesses" [All Fields] OR "assessing" [All Fields] OR "assessment" [All Fields] OR "assessment s" [All Fields] OR "assessments" [All Fields]) OR ("toma" [All Fields] AND ("drug effects" [MeSH Subheading] OR ("drug" [All Fields] AND "effects" [All Fields]) OR "drug effects" [All Fields] OR "de" [All Fields]) AND "datos" [All Fields]) OR (("registro" [All Fields] OR "registros" [All Fields]) AND ("drug effects" [MeSH Subheading] OR ("drug" [All Fields] AND "effects" [All Fields]) OR "drug effects" [All Fields] OR "de" [All Fields]) AND "datos" [All Fields]) OR ("recopilacion" [All Fields] AND ("drug effects" [MeSH Subheading] OR ("drug" [All Fields] AND "effects" [All Fields]) OR "drug effects" [All Fields] OR "de" [All Fields]) AND "datos" [All Fields]) OR "evaluacion" [All Fields]) **Translations** **music therapy:** "music therapy" [MeSH Terms] OR ("music" [All Fields] AND "therapy" [All Fields]) OR "music therapy" [All Fields] **technology:** "technology" [MeSH Terms] OR "technology" [All Fields] OR "technologies" [All Fields] OR "technology's" [All Fields] **technological:** "technologic" [All Fields] OR "technological" [All Fields] OR "technologically" [All Fields] **virtual:** "virtual" [All Fields] OR "virtuality" [All Fields] OR "virtualization" [All Fields] OR "virtualized" [All Fields] OR "virtualizing" [All Fields] OR "virtuals" [All Fields] **digital:** "digital" [All Fields] OR "digitalisation" [All Fields] OR "digitalised" [All Fields] OR "digitalization" [All Fields] OR "digitalize" [All Fields] OR "digitalized" [All Fields] OR "digitalizer" [All Fields] OR "digitalizing" [All Fields] OR "digitally" [All Fields] OR "digitals" [All Fields] OR "digitization" [All Fields] OR "digitizations" [All Fields] OR "digitize" [All Fields] OR "digitized" [All Fields] OR "digitizer" [All Fields] OR "digitizers" [All Fields] OR "digitizes" [All Fields] OR "digitizing" [All Fields] **tecnologia:** "tecnologia" [All Fields] OR "tecnologias" [All Fields] **data collection:** "data collection" [MeSH Terms] OR ("data" [All Fields] AND "collection" [All Fields]) OR "data collection" [All Fields] **assessment:** "assess" [All Fields] OR "assessed" [All Fields] OR "assessment" [All Fields] OR "assesses" [All Fields] OR "assessing" [All Fields] OR "assessment" [All Fields] OR "assessment's" [All Fields] OR "assessments" [All Fields] **registro:** "registro" [All Fields] OR "registros" [All Fields]

The update search yielded 252 articles from PubMed and 157 results from Scopus – 409 in total. After removing duplicate results – 39 articles – and collating the update search in both databases, 370 references remained.

After screening the 370 abstracts, 332 studies were excluded because they clearly reflected not using technology for data collection or assessment in music therapy. 38 studies appeared to fit these criteria through their abstract, so all these studies were read in full to check their suitability. Finally, 30 full-text articles were excluded, resulting in 8 included studies (see Fig. 1).

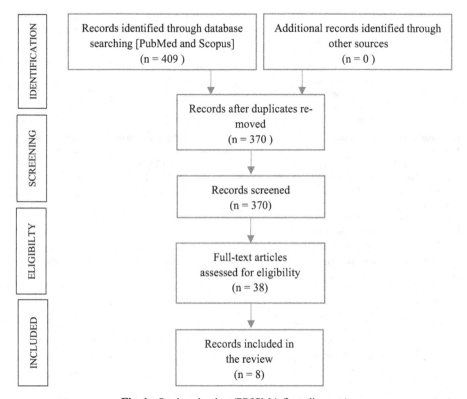

Fig. 1. Study selection (PRISMA flow diagram)

4   Results

Out of 370 studies, only 8 studies qualified our inclusion criteria: use technology for data collection or evaluation in music therapy. A detailed search criterion has been demonstrated in Fig. 1 and the details of the 8 articles are summarized in Table 2.

Table 2. Final selection.

Author/s	Research	Year of publication	Use of technology	Tools	Category
Lu D	Evaluation Model of Music Therapy's Auxiliary Effect on Mental Health Based on Artificial Intelligence Technology	2022	Assessment	Artificial Intelligence	Mental Health
Xu W	Response of music therapy to body mechanism and automatic efficacy evaluation system based on artificial intelligence technology	2022	Data collection and assessment	Artificial Intelligence	Emotions
Faramarzi A., Sharini H., Shanbehzadeh M., Pour MY., Fooladi M., Jalalvandi M., Amiri S., Kazemi-Arpanahi H	Anhedonia symptoms: The assessment of brain functional mechanism following music stimuli using functional magnetic resonance imaging	2022	Data collection	Magnetic resonance imagining SPM Toolbox in MATLAB software	Anhedonia and depression
Pascual-Vallejo C. D., Casillas-Martín, S., Cabezas-González, M	Una propuesta alternativa de evaluación en el diseño de las interacciones e interfaces tecnológicas musicales: en la búsqueda de nuevos paradigmas	2022	Assessment	IMTAP software	Music therapy
Alneyadi M., Drissi N., Almeqbaali M., Ouhbi S	Biofeedback-Based Connected Mental Health Interventions for Anxiety: Systematic Literature Review	2022	Data collection	Biofeedback devices	Anxiety

(continued)

Table 2. (*continued*)

Author/s	Research	Year of publication	Use of technology	Tools	Category
Luo Z., Durairaj P., Lau C.M., Katsumoto Y., Do E.Y.-L., Zainuddin A.S.B., Kawauchi K	Gamification of Upper Limb Virtual Rehabilitation in Post Stroke Elderly Using Silver Tune - A Multi-sensory Tactile Musical Assistive System	2021	Data collection and assessment	SilverTune	Rehabilitation in Post Stroke Elderly
Hatwar N. R., Gawande U. H	Can music therapy reduce human psychological stress: A review	2019	Data collection	Electroencephalogram (EEG)	Stress
Magrini M., Pieri G	Real time system for gesture tracking in psycho-motorial rehabilitation	2014	Data collection	Real-time gesture tracking	Psycho-motorial rehabilitation

5   Discussion

The 8 records included in this review show different ways of using technology for data collection and assessment in music therapy:

- Lu in 2022 [42] proposes an evaluation model of the auxiliary effect of music therapy on mental health based on artificial intelligence technology. For this, the research delves into big data mining theory and proposes a statistical data model, showing the applied formulas in the evaluation model and the rules for updating the information of the statistical data. The research uses an experimental group and a control group to evaluate its effectiveness and parameters such as pressure or stress are measured. The author concludes that "this model can effectively improve the accuracy of the evaluation of music therapy's auxiliary effect on mental health, and the evaluation results are accurate and reliable, with low error and good confidence" [42].
- A new system based on artificial intelligence is proposed in 2022 to evaluate the efficacy of music therapy. Xu [43] seeks to create a tool capable of monitoring the physical condition of subjects in real time and adjusting the type of treatment music according to this information. The proposal is mainly focused on the emotional aspect of people during music therapy sessions.
- Another recent study [44] used magnetic resonance imaging technology (fMRI) to determine the effect of music on depression and anhedonia. Three groups of people-healthy patients, patients with depression, and patients with depression and anhedonia-participated in the music therapy sessions, based on the stimulation with popular and traditional Iranian music. Then the data were analyzed using SPM Toolbox in MATLAB software. The authors highlight in conclusions that fMRI has been shown to be an effective tool in evaluating anhedonia symptoms by detecting the

hemodynamic changes in the cerebral cortex in response to appropriate musical stimuli [44].

– A research group in Spain has taken the tool IMTAP (The Individualized Music Therapy Assessment Profile) as a reference and they present in their article an alternative assessment proposal, highlighting the need to explore creative approaches to assessment in the design of cognitive, creative, and expressive technology [45]. It proposes an audiovisual record of the sessions - at least four cameras - and subsequent manual transcription of the data. In this way, the data is included manually, completing the categories established in the program form, and then an automatic analysis of the patient's evolution is made. In this way, the digital tool facilitates the assessment of the music therapy process.

– A 2022 systematic review by Alneyadi, Drissi, Almeqbaali and Ouhbi [46] evaluated interventions using technology for mental health care and biofeedback technology as a part of their process for anxiety management. The study reflects that biofeedback was used to better understand both psychological and physiological patient information - as well as the association between the two - during the intervention, but it also allows data collection that could be valuable for therapists to analyze the patient assessment.

– Luo, Durairaj, Lau, Katsumoto, Do, Zainuddin and Kawauchi [47] develop Silver-Tune, a smart multi-sensory musical assistive system to promote the rehabilitation outcomes of elderly post-stroke patients. The app allows users to work on their skills but also the app can record the user's profile and the data input from SilverTune for future analysis by the therapist. This automatic data collection - through app technology - facilitates a more accurate assessment of the patient's progress.

– Another project that uses technology to collect data on the effect of music on the brain of people with stress is the one developed by Hatear and Gawande [48]. They use electroencephalogram technology (EEG) as a non-invasive tool and to detect brain waves during the music.

– Finally, we found a system for real-time gesture tracking developed by the Signal and Images Lab of ISTI-CNR. Magrini and Pieri [49] affirm that the developed algorithms can extrapolate features from the patient - such spatial position, arms, and legs angles etc.- and the system the operator can link these features to sounds synthesized in real time, following a predefined schema. In this way, the movements made during the music therapy psycho-motorial rehabilitation sessions are automatically recorded.

Consequently, this systematic review shows that only 2,16% of the records referred to the technology available for data collection or for assessment in music therapy from 2013 to June 2023. However, more than half of these studies -5 out of 8- were published in 2022, showing a clear increase in interest nowadays in this theme.

One explanation for this current increase could be the need to have appropriate assessment tools in music therapy [23], adapted to the clinical context, to obtain significant and reliable results [50] and to assure of the quality of the evaluation for the continuous development of the discipline [17] But maybe it could be also triggered by the advent of artificial intelligence [43] and the new paradigm evidenced after COVID-19, which further demonstrates the need to find flexible and resilient technological designs in favor of health and well-being [44].

In contrast, 97.84% of records in this systematic review are focused on the use of technology with other functions; but not taking advantage of this technology for data collection or for subsequent evaluation. Among these excluded records, the most common use of technology is as part of music therapy sessions, either creating music, to promote the progress of the patient or helping achieve the music therapeutic goals. This group represents 25.41% of the records.

The 362 recorded excluded -for not meeting the inclusion criteria- are mainly divided into four themes:

- -The use of technology as part of music therapy sessions - either using digital instruments or devices that allow musical production.
- The technology applied to other therapies carried out by the patient, in a process that also combines traditional music therapy.
- Technology as part of the context and the day-a-day of the patient who attends music therapy.
- Technology as a tool to carry out music therapy sessions - due to distance or during the COVID-19 pandemic.

6   Conclusion

The findings of this systematic review clearly show that there has been great growth in the use of technology in music therapy in the last 10 years, but this has been focused on technology as a complement to carrying out music therapy sessions or using technology systems to facilitate musical production. However, there is growing the number of studies focused on finding more accurate solutions for data collection and assessment in music therapy through technology.

A special growth is registered in 2022 and it seems that music therapy data collection and assessment technology could have more development soon. The advent of artificial intelligence and the consequences of the COVID-19 pandemic may have motivated this increase, but this study suggests that the research in this context is insufficient for this moment, so it is necessary to continue looking for technological tools for data collection and assessment in music therapy to ensure the quality and achieve a standardized process in the future.

References

1. Wigram, T., Gold, C.: Music therapy in the assessment and treatment of autistic spectrum disorder: Clinical application and research evidence. Child Adolesc. Mental HealthAdolesc. Mental Health 17(2), 52–61 (2012). https://doi.org/10.1111/j.1475-3588.2011.00628.x
2. Dileo, C., Bradt, J.: Medical music therapy. In: Edwards, J. (ed.) The Oxford Handbook of Music Therapy, pp. 366–380. Oxford University Press, Oxford (2013)
3. Stige, B.: Participation, agency, and change in music therapy: reflections on a qualitative case study in dementia care. Nord. J. Music. Ther.Ther. 25(2), 154–172 (2016). https://doi.org/10. 1080/08098131.2015.1059479
4. Pavlicevic, M.: Standardizing evaluation in community music therapy: challenges and possibilities. Voices: World Forum Music Ther. 18(2) (2018). https://voices.no/index.php/voices/ article/view/948/777

5. O'Kelly, J., O'Callaghan, C.: Music therapy outcome measurement: the importance of alignment with music therapy theory and practice. J. Music Ther.Ther. **57**(3), 223–245 (2020). https://doi.org/10.1093/jmt/thaa011

6. Magee, W.L., Davidson, J.W., Hwang, Y.: Constructing a taxonomy of the qualities of music therapy interactions in cancer care. J. Music Ther.Ther. **52**(4), 514–546 (2015). https://doi.org/10.1093/jmt/thv010

7. Mössler, K., Chen, X.J., Heldal, T.O., Gold, C.: Music therapy for people with schizophrenia and schizophrenia-like disorders. Cochrane Database System. Rev. **11**, CD004025 (2018). https://doi.org/10.1002/14651858.CD004025.pub4

8. Schmid, W., Ostermann, T., Särkämö, T.: Neural correlates of improvisation in music therapy: an activation likelihood estimation meta-analysis. Front. Hum. Neurosci.Neurosci. **13**, 462 (2019). https://doi.org/10.3389/fnhum.2019.00462

9. Magee, W.L.: The art of research and the research of art: charting a course for music therapy. J. Music Ther.Ther. **58**(1), 1–7 (2021). https://doi.org/10.1093/jmt/thaa034

10. Wheeler, B.L.: Evidence-based practice in music therapy: from theory to practice. Music. Ther. Perspect.Ther. Perspect. **35**(2), 111–118 (2017). https://doi.org/10.1093/mtp/mix013

11. O'Callaghan, C., Baron, S.: Revisiting standardized assessments: Integrating contemporary perspectives on assessment and evidence-based practice. Nord. J. Music. Ther.Ther. **27**(1), 70–88 (2018). https://doi.org/10.1080/08098131.2017.1298276

12. Schmid, W., Elefant, C.: Outcome research in music therapy: a selective review of the current literature. Front. Psychol. **11**, 2520 (2020). https://doi.org/10.3389/fpsyg.2020.571321

13. Gattino, G.S.: Essentials of music therapy assessment. Forma e Conteúdo Comunicação Integrada (2021)

14. O'Kelly, J., Magee, W.L.: The use of standardized outcome measures in music therapy research: a literature synthesis. J. Music Ther.Ther. **52**(3), 324–359 (2015). https://doi.org/10.1093/jmt/thv006

15. Gattino, G.S.; Developing and adapting music therapy assessment tools. In: AMTA 2020 Abstract Book Australian Music Therapy Association Inc. (2020)

16. Schmid, W., Ostermann, T., Elefant, C., Stige, B.: Towards a common terminology: a simplified framework of interventions and treatment goals in music therapy. Nord. J. Music. Ther.Ther. **23**(1), 71–90 (2014). https://doi.org/10.1080/08098131.2013.828173

17. Edwards, J.: Evaluating music therapy practice with a quality assurance framework. Nord. J. Music. Ther.Ther. **25**(3), 246–264 (2016). https://doi.org/10.1080/08098131.2016.1172313

18. Sotelo, J.A.: Ojo al dato. Cada registro es un valioso tesoro. I Jornada para el impulso profesional del musicoterapeuta UNIR. Madrid, Spain (2023)

19. Tsiris, G.: La investigación y buena praxis al final de la vida. I Simposium Internacional de investigación y buena praxis en Musicoterapia SOCIEMT. Plasencia, Spain (2022)

20. Benenzon, R.: Music therapy training: the contribution of scientific research to clinical practice. Music Med. **6**(1), 19–22 (2014). https://doi.org/10.1177/1943862113518491

21. Magee, W.L., Burland, K.: Music therapy research and practice in healthcare contexts: a critical commentary. Front. Psychol. **11**, 611023 (2020). https://doi.org/10.3389/fpsyg.2020.611023

22. Gattino, G.S.: Avances en la investigación de la evaluación en Musicoterapia. I Simposium Internacional de investigación y buena praxis en Musicoterapia SOCIEMT. Plasencia, Spain (2022)

23. Waldon, E.G., Jacobsen, S.L., Gattino, G.S.: Music Therapy Assessment: Theory, Research, and Application. Jessica Kingsley Publishers, London (2018)

24. Wosch, T., Wigram, T.: Microanalysis in music therapy: methods and techniques for analyzing the therapeutic process. J. Music Ther.Ther. **51**(4), 292–320 (2014). https://doi.org/10.1093/jmt/thu036

25. Holk, U.: An ethnographic descriptive approach to video microanalysis microanalysis in music therapy. In: Wosch, T., Wigram, T. (eds.) Microanalysis in Music Therapy: Methods, Techniques and Applications for Clinicians, Researches, Educators and Students. Jessica Kingsley Publishers, London (2014)

26. Plahl, C.: Microanalysis of preverbal communication in music therapy. In: Wosch, T., Wigram, T. (eds.) Microanalysis in music therapy: Methods, Techniques and Applications for Clinicians, Researches, Educators and Students. Jessica Kingsley Publishers, London (2014)

27. Ridder, H.M.: Microanalysis on selected video clips with focus on communicative response in music therapy. In: Wosch, T., Wigram, T. (eds.) Microanalysis in Music Therapy: Methods, Techniques and Applications for Clinicians, Researches, Educators and Students. Jessica Kingsley Publishers, London (2014)

28. Scholtz, J., Voigt, M., Wosch, T.: Microanalysis of interaction in music therapy (MIMT) with children with developmental disorders. In: Wosch, T., Wigram, T. (eds.) Microanalysis in Music Therapy: Methods, Techniques and Applications for Clinicians, Researches, Educators and Students. Jessica Kingsley Publishers, London (2014)

29. Wigram, T.: Event-based analysis of improvisations using the improvisation assessment profiles (IAPs). In: Wosch, T., Wigram, T. (eds.) Microanalysis in Music Therapy: Methods, Techniques and Applications for Clinicians, Researches, Educators and Students. Jessica Kingsley Publishers, London (2014)

30. Baker, F.: Using voice analysis software to analyze the sung and spoken voice. In: Wosch, T., Wigram, T. (eds.) Microanalysis in music therapy: Methods, Techniques and Applications for Clinicians, Researches, Educators and Students. Jessica Kingsley Publishers, London (2014)

31. Abrams, B.: The use of improvisation assessment profiles (IAPs) and RepGrid in microanalysis of clinical music improvisation. In: Wosch, T., Wigram, T. (eds.) Microanalysis in music therapy: Methods, Techniques and Applications for Clinicians, Researches, Educators and Students. Jessica Kingsley Publishers, London (2014)

32. Sutton, J.: The use of micro-musical analysis and conversation analysis of improvisation: 'the invisible handshake' – free musical improvisation as conversation. In: Wosch, T., Wigram, T. (eds.) Microanalysis in Music Therapy: Methods, Techniques and Applications for Clinicians, Researches, Educators and Students. Jessica Kingsley Publishers, London (2014)

33. Trondalen, G.: A phenomenologically inspired approach to microanalyses of improvisation in music therapy. In: Wosch, T., Wigram, T. (eds.) Microanalysis in Music Therapy: Methods, Techniques and Applications for Clinicians, Researches, Educators and Students. Jessica Kingsley Publishers, London (2014)

34. Pavlicevic, M.: The music interaction rating Scale (Schizophrenia) (MIR(S)) microanalysis of co-improvisation in music therapy with adults suffering from chronic schizophrenia. In: Wosch, T., Wigram, T. (eds.) Microanalysis in Music Therapy: Methods, Techniques and Applications for Clinicians, Researches, Educators and Students Jessica Kingsley Publishers, London (2014)

35. Inselmann, U.: Microanalysis of emotional experience and interaction in single sequences of active improvisatory music therapy. In: Wosch, T., Wigram, T. (eds.) Microanalysis in music therapy: Methods, Techniques and Applications for Clinicians, Researches, Educators and Students Jessica Kingsley Publishers, London (2014)

36. Stige, B., Ansdell, G., Elefant, C., Pavlicevic, M.: Where Music Helps: Community Music Therapy in Action and Reflection. Ashgate Publishing, Farnham (2017)

37. Ruud, E.: Music Therapy: A Perspective from the Humanities. Barcelona Publishers, New Braunfels (2010)

38. Knight, A.: Uses of iPad® applications in music therapy. Music. Ther. Perspect.Ther. Perspect. **31**(2), 189–196 (2013). https://doi.org/10.1093/mtp/31.2.189

39. Baxter, H.T., Berghofer, J.A., MacEwan, L., Nelson, J., Peters, K., Roberts, P.: The Individualized Music Therapy Assessment Profile, IMTAP. Jessica Kingsley Publishers, London (2007)
40. Chau, R., Lamont, A.: Virtual Music Instrument Homepage, https://hollandbloorview.fli ntbox.com/technologies/fdd00282-c837-4108-ad10-102b0a64da2d. Accessed 07 July 2023
41. SoundBeam Homepage. https://www.soundbeam.co.uk/. Accessed 07 July 2023
42. Lu, D.: Evaluation model of music therapy's auxiliary effect on mental health based on artificial intelligence technology. J. Environ. Publ. Health **9960589** (2022). https://doi.org/10.1155/2022/9960589
43. Xu, W.: Response of music therapy to body mechanism and automatic efficacy evaluation system based on artificial intelligence technology, In: 2021 2nd International Seminar on Artificial Intelligence, Networking and Information Technology (AINIT), Shanghai, China, pp. 57–60 (2021). https://doi.org/10.1109/AINIT54228.2021.00020
44. Faramarzi, A., et al.: Anhedonia symptoms: the assessment of brain functional mechanism following music stimuli using functional magnetic resonance imaging. Psychiatry Res. Neuroimaging 111532 (2022). https://doi.org/10.1016/j.pscychresns.2022
45. Pascual-Vallejo, C.D., Casillas-Martín, S., Cabezas-González, M.: An alternative evaluation proposal in the design of musical technological interactions and interfaces: looking for new paradigms. Artseduca **33**, 152–165 (2022). https://doi.org/10.6035/artseduca.6896
46. Alneyadi, M., Drissi, N., Almeqbaali, M., Ouhbi, S.: Biofeedback-based connected mental health interventions for anxiety: systematic literature review. JMIR Mhealth UhealthMhealth Uhealth **9**(4), e26038 (2021). https://doi.org/10.2196/26038
47. Luo, Z., et al.: Gamification of upper limb virtual rehabilitation in post stroke elderly using SilverTune- a multi-sensory tactile musical assistive system. In: IEEE 7th International Conference on Virtual Reality (ICVR), Foshan, China, pp. 149–155 (2021). https://doi.org/10.1109/ICVR51878.2021.9483850
48. Hatwar, N.R., Gawande, U.H.: Can music therapy reduce human psychological stress: a review (2020). https://doi.org/10.1007/978-981-15-0077-0_41
49. Magrini, M., Pieri, G.: Real time system for gesture tracking in psycho-motorial rehabilitation. In: HEALTHINF 2014 - 7th International Conference on Health Informatics, Proceedings; Part of 7th International Joint Conference on Biomedical Engineering Systems and Technologies, BIOSTEC 2014, pp. 563–568 (2014). https://doi.org/10.5220/0004937205630568
50. Magee, W.L., Baker, F.A.: The role of embodied listening in promoting the development of a therapeutic relationship in music therapy. Nord. J. Music. Ther.Ther. **27**(2), 123–144 (2018). https://doi.org/10.1080/08098131.2017.1311407

Network Dance and Technology

Sound Body as Embodied Poetic Interaction

Gatti Daniela$^{(\boxtimes)}$ (iD)

Universidade Estadual de Campinas (UNICAMP), São Paulo, Brazil
danigati@unicamp.br

Abstract. The Nucleo de Dança Redes at the State University of Campinas has been developing multimodal artistic research in mixed reality in an interactive, immersive, dialogical, and in nets. In this article we present some research focus of the group referencing two works "De uma Margem a Outra" and "Jardim das Cartas". Researchers in the areas of dance, music and audio visual have been raising discussions about an "expanded sound body" that stems from the interaction between movement, sound, music, video and technology in a mixed environment. A body that extends in sound, and an embodied sound in movement in an expanded perspective of the body in "presence" with more complex contours.

Another point is in the observation and understanding of the event of multimodal improvisation in a mixed environment as a path and an interactive and connective articulation between the communication agent that participate in the net as an expressive path that is shown in the interactive exploration. Based on interactive practices and on the concept of net, we work methodologically towards a "poetic interaction" in the field of complexity. A fertile ground in which improvisation and composition become coordinated paths for the unfolding of a dance which starts from the connectivity and from the interaction between sound, body movement, image and technology.

Keywords: sound body embodied · interactivity · improvisation · multimodality

1 Sound Body: Extended Movement in Sound/Embodied Sound

Nucleo de Dança Redes[1] develops research which investigates the body and movement in creative processes in interaction with other languages and realities. Several artistic works were carried out in partnership with researchers in the areas of arts, especially in

---

[1] Núcleo de Dança Redes research group was created in 2012 by Associate Professor and researcher Daniela Gatti in Corporal arts Department of the Institute of Arts at the State University of Campinas BR. it is accredited by the CNPQ (National Research Center) linked to the graduate school program in Performing Arts. Formed by researchers in the areas of dance and performance with the objective of developing multidisciplinary artistic research and creative processes in nets integrating new digital technologies in the field of performing arts with emphasis on studies focused on improvisational and compositional processes in the field of complexity. Since 2012 the research group has already produced more than 40 artistic works and carried out scientific research in the field of creativity, interdisciplinarity and education.

© ICST Institute for Computer Sciences, Social Informatics and Telecommunications Engineering 2024
Published by Springer Nature Switzerland AG 2024. All Rights Reserved
A. L. Brooks (Ed.): ArtsIT 2023, LNICST 564, pp. 57–72, 2024.
https://doi.org/10.1007/978-3-031-55319-6_5

the sound and musical field and in the multimodality with composers. Since 2013 the partnership between Gatti and Professor Jonatas Manzolli[2] has explored new poetic and aesthetic perspectives in the interaction between dance, sound language, multimodal, musical, computing and technological language.

It is worth mentioning that collaborations between artist composers and choreographers have been present in the western artistic historical context. With disruptive, innovative and avant-garde visions with an interest in investigating convergences and divergences in the forms and structures, practices and aesthetics of each language. Some partnerships between music and dance from the 20th century onwards were marked by experimentalism from a relational and interactive perspective like, e.g., George Balanchine and Igor Stravinsky, Martha Graham and Louis Horst, John Cage and Merce Cunningham; Thierry De Mey and Anne Teresa de Keersmaeker, which broke paradigms in compositional processes and in creation.

Highlighted in this article we rescue two artistic projects carried out as a result of the academic and artistic partnership in order to explore dance and sound integrated into computational systems and net from mediation by interfaces and use of accelerometers, sensors attached to the bodies of the dancers and also in their relationship with mobile phone devices. Both experiences followed a rhizomatic perspective in expecting the body to become extended in sound; and the embodied sound "embodied" in "sound-body".

Sound Body concept can also be seen as Extended Body in sound or emboided. The idea of sound body merges with the term embodied which means an expression or giving tangible or visible form to (an idea, quality or feeling). Which can be seen as a body that extends to another condition:

1. To give a bodily form to; incarnate;
2. To represent in bodily or material form;
3. To make part of a system or whole; incorporate.

The embodied term presents its complexity as it permeates different areas of philosophy, psychology and sociology, with its specific meaning depending on the specific discipline. From a phenomenological perspective, the body is seen as the center of identity, inseparable from sensory experience and perception. Incorporation generally happens in and with the body and its interactive processes, which through the senses, assist, improve or interfere with human development. In the context of arts and multimodality, the emphasis is on the relationship between the experience of the body and

---

[2] Jonatas Manzolli is a full professor in the department of music at the Institute of Arts at the State University of Campinas. Researcher at CNPq and member of the Executive Board of the Interdisciplinary Sound Communication Nucleus (NICS). He performs as a Professor of composition subjects, creation with new supports and orchestration in the Music Department (IA/UNICAMP). In addition to these activities, he has worked throughout his career in international institutions, from his PhD at University of Nottingham (1989-1993), UK and his studies at the Institute of Sonology (1991-92), The Netherlands. He was also a guest researcher at the Neuroinformatics Institute of ETHZ, Switzerland, (1998-2004) and at the SPECS group at Pompeu Fabra University, Spain, (2005-2015). He is currently a researcher and collaborator of the Center for Interdisciplinary Research in Music and Technology (CIRMMT) at the University McGill, Canada (since 2015) and INET- MD Institute of Ethnomusicology - Center for Studies in Music and Dance in Portugal. https://linktr.ee/jotamanzo.

multimodal resources, practices, media and social spaces and this entire relationship is interdependent, In the area of music for example, Leman [1] presents research on the relationship between body, movement and sound based on the idea of emboidy.

Embodied music interaction manifests itself through activities with sounds (listening, playing, dancing), with other people (as in joint action), as well as with music instruments and within the body (as a mediator for music playing). The interactions are constrained, though, by acoustical structures (both in music and in the radiation of sounds), by cognitive activities (limitations of memory, attention, learning), and by body resonances, biomechanical, and metabolic and energetic restrictions. Many authors believe that musical constraints can be better understood by considering the timing of embodied interactions, such as the rhythmic coordination of the human body with external musical rhythms, given the nature of music as a temporal art form. This field of study started with experiments in which subjects were asked to tap their finger along with metronomes and with music. [1].

When it comes to sound body, the body merges with sound in an expanded sensorial experience. Le Breton [2] emphasizes the interaction of the body's senses in relation to its perception of the world. That is, this relationship naturally tends to break down boundaries in a correspondence existing between the body and the sound. One acts on the other in a kind of sensorial web in resonance, where the stimulations can happen in correspondence. The body's senses interact with the sound universe in a relational action.

From an anthropological perspective, the work Sound in embodied practice [15], corroborates the idea that every sensation experienced from sound is always incorporated, considering the vibrational idea and sound waves as a stimulus for affecting the entire body, and not just the ears. The authors consider sound not only as an interpretative approach but based on experience and multisensory sound study.

We find aspects of embodied cognition that emphasize interaction as an entanglement of bodily, environmental, and social components as embodied, extended, and active components of mind and cognition. [16].

Leman presents the paradigm of embodied music cognition provides an interesting viewpoint on musical gestures. It assumes that music is based on a tight relationship between sounds and experiences that are mediated by the body [17] (see Fig. 1).

The paradigm of embodied musical cognition for exemple, is based in a series of concepts, related to: (i) the body as mediator, (ii) the gesture/action repertoire, (iii) the action-perception coupling and (iv) the link with the subjective experiences, such as intentions, expressions, empathy, and emotions. [17].

We can consider in the artistic works presented in this article a conceptual similarity with regard to poetic production that arises from emboided at the intersection between sound, body, image movement and technology as dynamic and complex system.

1.1 Network Poetry - De uma Margem a Outra

The multimodal performance "De uma Margem a Outra"[3] created in 2018 was based on, for investigation, the idea of "body" as a place of sound passage and displacement. The multimodal work had its agents "body, text, voice, metaflute, percussion, electroacoustic music, technological devices and pendular objects, integrated as a diagram, in an articulation of several elements in con(fluency) for a "dynamic musicoreographic" structure (Fig. 2).

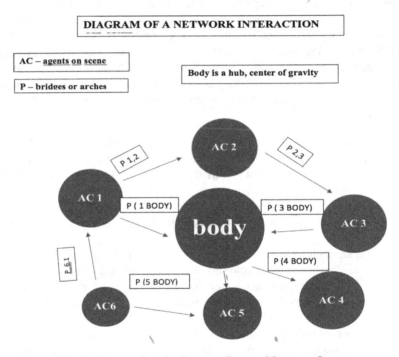

Fig. 1. Diagram Creative Process - De uma Margem a Outra

In the performance "De uma Margem a Outra", body, sound, voice and musical instruments were interconnected by devices and transducers (sensors and microphones attached to the musical instruments, to the dancers and the voice) where the signals were sent and processed by computational models created to establish the interactive environment. The dancers' bodies were receivers of a circuit articulated in a network in which sound was generated by movement through sensors attached both to the dancers' bodies

---

[3] Dance, music and technology multimodal performance created in 2017 and 2018 inspired by the literary work "Six proposals for the next millennium" by Italo Calvino with funding from the Culture Department of São Paulo State PROAC 2017. Directed by Daniela Gatti and Jonatas Manzolli. Dancers - Karina de Almeida and Tutu Morasi; Percussionists GRUPU (Percussion Group of Unicamp): Rodolpho Simmel, Otavio Antoniecci and Rafael Peregrini; Flutist - Gabriel Rimoldi; Laiana Oliveira – Mezzo Soprano; Danilo Rosseti – Electroacoustic Composer and Operator and software processing.

Fig. 2. De uma Margem a Outra - Karina Almeida (dancer); Gabriel Rimoldi (metaflute) Foto: Kassius Trindade

and to musical instruments (metaflute, percussion and voice) and also to scenic objects in pendulum movement in space. The musical and choreographic composition was designed from the perspective of movement being the passage to sound representation as an extended body and also in a poetics of interaction (Fig. 3).

The body research covered in its extension in sound was managed by the computational interference inducing the body of the dancers to explore the movements in a new and expanded territory regarding relations and attitudinal aspects.

Stimulating the exercise of interaction led the dancers to a more expanded perception on movements when connected to sound. An attentional relationship pointed:

Fig. 3. De uma Margem a Outra - Tutu Morasi (dancer); Laiana Oliveira (singer) and percussion. Foto: Kassius Trindade

- to the recognition and perception of a body in transit and dialogue in action (how to improvise in sound and net interaction).
- to the structures of the language itself (dynamics, directions, displacements, activation of body parts, rhythm).
- to the recognition of new models and gestural protocols triggered and recurrent during improvisation.
- to decision-making between acting, reacting and pausing in actions or situations - in the perception of convergences and divergences between the agents.

Such points were observed and raised during the experimental research in which we realized that musicians, dancers and a net computing environment built a dynamic and complex self-organizing system sharing their collective and own visions and practices. Thus, the complexity resides precisely in this arena in which all of them create realities together in a variability that demands adjustments, changes of routes, of choices and abandonment of ideas.

"De uma Margem a Outra" awakened to a self-organizing multimodal methodology from the interactive improvisational and compositional practice between dance, music, text and sound in more complex environments and integrated into multimodal and computational systems. A practice that develops in the anticipatory experience of the action and in the action itself, from a more expanded perception of the exercise of interaction. It is in action that integration and collaboration take place.

In line with the idea of expanded perception, Nhur [3] in her article Do Movimento ao Som, Do Som ao Movimento: relações bioculturais entre dança e música, (From Movement to Sound, From Sound to Movement: biocultural relations between dance and music), brings important aspects about the levels of interaction which happen between sound and movement referencing the author Marc Leman[4] [4] in his reflections. The author analyzes two dance performances "Peças Fáceis" by the theater group Pró –Posição and "Z" by Alejandro Ahmed from Cia Cena 11 in which both integrate music and sound to conceptions of embodiment in a socio-environmental and cultural perspective. Nhur [3] highlights that the synchrony between action and perception rests on the hypothesis of body intentionality and points out the indissociability between perception and action, being considered as the responsible agent in the communication between sound and movement. And adds that.

> [...] the way an organism receives the sound energies of the environment and turns them into abstract actions or concepts depends on an ecological perspective, mediated by the resonance between natural and cultural domains. [3].[5]

We can consider that "extended" and/or "expanded" perception is always accessible to everyone as long as the agents are in a state of acute attention turning to the relationships and that are present in what is in the "between body". That is to say, the place of the sound is in the movement itself and in the net simultaneously. A job that investigates the dialectic of action and reaction between sound, gesture and silence in an expanded space and in a mutual and complex correspondence.

Experiences from the creative process "De uma Margem a Outra" brought reflections on the body that becomes sound, movement, in net as a living hybrid organism. A dance that is reflected in a body which appropriates vibrations and sound waves, dissonances and harmonies, sound textures and pulsations. A body that stretches and unfolds from the "interaction" with itself, the sound, the movement and every technological apparatus with its supports and systems which mediate traveling in a new poetic path.

1.2 Extended Body

The term Extended Body refers to a new interpretative and poetic approach in dance borrowing from the musical sound field the idea of "extended technique" developed to musical instruments and voice.

At the beginning of the 20th century, musical composers began to develop and expand the vision of tonality seeking new horizons of possibilities on how to deal with the sound [5]. New compositional currents appeared such as serialism, twelve-tone music, atonalism, electronic music, random music, electroacoustic music, concrete music, among others. These compositional currents in Europe and the United States consolidated the

---

[4] Marc Leman is Methusalem research professor in systematic musicology and director of IPEM, the institute for psychoacoustics and electronic music at Ghent University. He holds MA degrees in musicology and philosophy, and did his PhD on computer modeling of tonality perception book (Embodied Music Cognition).

[5] Original text in portuguese: [...]o modo como um organismo recebe as energias sonoras do ambiente e as transforma em ações ou conceitos abstratos depende de uma perspectiva ecológica, mediada pela ressonância entre os domínios do natural e do cultural" [3].

idea of experimentalism which marked the 20th century with the arts. "Extended Techniques" came to be used in reference to instrumental exploration techniques in the search for new sounds" [6].

The voice also resorts to these techniques in the investigation of vocal resources in gestures that break the tradition of singing by including in the repertoire: screams; whispers; smiles; babbling; spoken voice; narrated voice; whispered voice; morphings; among many other ways of vocal emission [7].

The proposal to break with traditional models in the sound field provoked an expansion and potentiation of the elements and in the acoustic characteristics of the sound such as intensity, frequency, duration, sound wave, besides the timbre possibilities involved in the sound field.

Padovani and Ferraz [6] state that extended techniques in the musical field are those which are unusual within a given historical, social and aesthetic context.

Thus, borrowing the term extended technique used in the musical and sound area for dance may be a start for understanding an "extended body" as rupture movement to the sensory and perceptive field in the relationship of the body which dances with digital technology and different environments in expanded reality such as virtual reality [VR] mixed reality [MR] expanded reality [AR].

An extended body interacting with other environments can change paradigms already traditionally preconceived in relation to the movement that works with time and space in its biomechanical configuration. The body which dances in virtuality and in connection with technological and digital devices runs through other registers yet to be explored in their relationship with gravity, speed, simultaneity, duplication of bodies, non-physical presence, extension, etc. The hypothesis is that digital technology participates and is part of the extended body.

The idea of an extended body still requires studies in the field of body language and dance in the dimensions of virtual, mixed or expanded environments and of structuring elements of the language of movement itself. Appropriating the term extended technique to extended body can be a fertile field for exploration in the relationship of new poetics and perceptions of the movement.

1.3 Multimodal Installation *Jardim das Cartas*[6]

In this same investigative line that pursues the scope of complexity in dance in multimodality with interaction and computational processing, we proposed in 2022. The multimodal installation in mixed reality "Jardim das Cartas" developed through the project Poetics of sound and body in the interdisciplinary dialogue between art and technology: a research study of creation in Digital Humanities, boosting research in the interactive perspective. According to Milton Sogabe [8] Interactive installations have the necessary elements to characterize the space of interactivity being environment, public, interfaces, digital manager and devices. "In addition to physical elements, there are processes that take place in time: event, interaction and information processing with input and output signals.

---

[6] Jardim das Cartas official channel: https://www.youtube.com/channel/UCy2WzV0o-Viw_tFk O74pL_A 2; Jardim das Cartas official website: https://linktr.ee/jardimdascartasoficial.

"Jardim das Cartas" was conceived from the beginning as an open work and brought together a complex information net in mixed reality with elements of dance, music, poetry, videoprojection sound, mobile phone devices, applications (Move Guitar), computational processing (Puredata) and database with 22 videos and voice recordings and software.

The work created virtual, face-to-face and telematic interactions where interactive liveperformances were proposed considering different audience environments.

The system, consisting of two programs implemented in Pure Data (Pd), a visual programming language developed by Pucket and TouchDesigner computer graphics platform, generates music and video in real time, respectively. Pure Data (Pd), widely used by musicians, and sound artists, is open-source project with a large developer base working on new extensions.

The TouchDesigner, developed by the Toronto-based company Derivative, and also used in Generative Art, is a node- based visual programming language for generating real-time interactive visuals. In live performance, these two programs allow the real-time interaction of audiovisuals with musicians and dancers and control a mixed-reality environment for virtual, presential, and telematic interactions [9] (Fig. 4).

Fig. 4. Live performance installation Jardim das Cartas 2022 (improvisation– sax; dancers with mobile phone, video, and piano improvisation online, and computer sound)

The Jardim das Cartas multimodal installation promoted a cooperative creative arena in mixed reality. Immersive agents in both physical and digital/virtual environments explored interactions through improvisation. A dynamic generative installation with possibilities of generating a corporeality from different divergent and convergent patterns. It is important to highlight that interactions in a cooperative state (multisensory space) could result in more dynamic responses to the stimuli received (see Fig. 5).

[…] Theorists of embodied cognition have already argued that one of the three dimensions of corporeality is intersubjective interaction (along with what they call "bodily self-embodiment"). Regulation" and "sensorimotor coupling"). Earlier constitutive "extended" features of the embodied structure, exploring the multisensory peripersonal space of musicians following a cooperative/non-cooperative (jazz) musical interaction [16] (Fig. 6).

Fig. 5. Live performance installation Jardim das Cartas 2022

Articles published by the researchers on the work Jardim das Cartas performance aspects of the methodological and theoretical development of the process with reflections that involve the conceptual and aesthetic character of the work as for the perception of a "ecological presence" [10] that integrates natures in mixed environments. And develop the entire conception of presence based on Gibson's affordances.

The concept of affordance is connected to the notion that Gibson constructs about "sense" or "feeling" in "The Senses Considered as Perceptual Systems" (Gibson, 1968). Considering the senses as perceptual systems is to say that the action of feeling is identical to perceiving. Perception cannot be distinguished from sensations derived from the act of interacting with the external world through the senses. The environment, then, plays a central role in the perceptual process. [10].

Fig. 6. Live performance installation Jardim das Cartas 2022 (dancers, piano online, video and sound computer online)

Presence is also a key-concept in the study of teleoperation, virtual reality (VR), virtualization of acoustic sources, games and image interaction. Since the early 1980s, a great effort of research focuses on establishing the parameters that rule the perception of Presence. For example, it has been found that detailed visual scenes are not so important, while multimodal convergence, representations of the body itself and environmental engagement, from sound and visual sources, increase the reported Sense of Presence [11].

The methodological assumption of both works placed the interactive practice as central in the research, improvisation and the self-organized process as a reflective and theoretical framework, where the reading of experiences in dynamic systems and the interweaving between agents produced a unique writing in the global, multimodal and interactive dimension between the different modes of communication in a complex rhizomatic perspective and in net.

It should be noted that one of the points of interest in net investigations with dance and other agents in mixed reality is the creative path that is unique, not constituting as a model to be followed and replicated but as a fluidic, dynamic and open methodology. The path problematizes procedural and systemic work of creation on the net opening up space for poetic interaction.

In 2023 in order to consolidate the methodological process of "Jardim das Cartas" we propose the ongoing project Expanded Sound Body (ESB): expanded spaces between sound, movement and technology in mixed reality[7] to direct the focus on the event of improvisation from the sound-body perspective. In this research we also enter the sphere

---

[7] Expanded Sound Body (ESB): expanded spaces between sound, movement and technology in mixed reality - project financed by FAPESP process -number - 2022/05935-7 researchers:Daniela Gatti; Jonatas Manzolli, Mariana Baruco Andraus; Alexandre Zamith

of multimodal creation, in which the assumption made is that the interaction of an agent or group of agents with an immersive space, using several interactive devices, indicates how these processes affect their expressive and poetic behavior and the meaning which is constructed by them.

The notion of interaction with which we work goes beyond the analytical dimensions only and will be linked to various perceptual modalities. Interactive narrative is seen as a result of processes, specifically from the interaction with all the agents that participate in the experience [11].

2 Improvisation in Mixed Reality and the Expanded Body in Presences

Interacting and improvising in dance are conjoined and complementary actions, as both reflect the same state of suspension that waits, listens and awaits the emergence of the apparent impulse of the action, movement and gesture. In this way, improvising and interacting are the perfect vias for creativity.

The term interact is a verb which derives from the noun interaction. It is a word which is formed by the prefix inter (meaning between) + noun action, i.e., it is characterized by an integrated action. Interacting is the same as being in mutual exercise with something, promoting changes in the development and/or state of both. It is in this dialogical context between action and non-action that we resort to improvisation as the event that mediates, connects and creates bridges for interactivity.

Improvise, which comes from the Latin IN PROMPTU, means: "in a state of attention, ready to act", "in readiness", from PROMERE, "to make arise". The meaning of the word readiness is "to be available at any time, circumstance and environment to make decisions".

The field of improvisation is per se the way of work that it proposes in its scope of investigation, the alternation of states of predictability and unpredictability, in random and autonomous processes. The agents who live the improvisational experience, reposition themselves at every moment of the action in a circularity between organization and chaos.

> Improvisation, as a singular expression within the field of contemporary dance, presents itself as a web of affections produced by the physical forces acting on bodies. It is a difficult experience to categorize within the current social representations, as it does not refer to itself as a "logocentrism". Its transmission does not meet criteria for ordering and systematization; on the contrary, it presents itself as a way of experimentation: [12].[8]

Almeida; Manuel Falleiros; Diogo Angeli; Elena Partesotti; Vania Eger Pontes; Guilherme Zanchetta Lalier Avila; Andrea Albergaria.

[8] Original text in portuguese: "A improvisação, como expressão singular dentro do campo da dança contemporânea, apresenta-se como uma trama de afetos produzida pelas forças físicas atuando sobre os corpos. Constitui uma experiência difícil de categorizar dentro das representações sociais vigentes, pois não se refere a si mesma como um "logocentrismo". Sua transmissão não responde a critérios de ordenamento e sistematização; pelo contrário, se apresenta como uma forma de experimentação: [HARISPE, 2014: 35].

Improvisation happens in synergy with known and unknown elements, which allows them to be awakened and apparent, creating in this hybrid space of what is known and not known in informational sources, forms and materials, becoming fabric and texture for both the observer and the improvising agent.

In the sphere of multimodality in mixed reality, there is a multi-connective transmission, in net and dialogical among all the performative agents and languages which become connected during the improvisational event, including several coexisting communicational modes and expanded reality that may or may not be working simultaneously. It should be noted that what we call reality is not restricted to the concrete physical field, but reality as "truth while existence"; also is and happens in the mixed environment.

From meetings, failures to meet, alternations and suspensions, the event of improvisation lies on its own space of unpredictability and appears resulting from its own aesthetics. Improvisation in multimodality only happens with agents available to reconnect the communication where the process becomes autonomous while a state of presence.

In improvisational action in a mixed environment, the aim is to find embodied new strategies and methodological paths inferring in the aesthetic field a new perspective from the connections between reality and virtuality and implicit and explicit models that shuffle, suspend, cut, unite, juxtapose, alter, and separate communicative modes.

The game between virtuality and reality is complementary, as Levy [13] points out that the virtual is not opposed to the real, but to the current. Virtuality and reality are just two ways of being many different. The possible is already constituted and will be realized without that nothing changes in its determination or its nature. It is a latent real:

Multimodal improvisation is the act of dialoguing with, creating, reinventing or (trans)forming something in the present moment of realization, that is, it is an act of creation of open share. One of the characteristics of improvisation in the "arts of presence"-traditionally nominated for the performing arts (dance, theater and performance), is the sharing in its quality of the interaction between the performative agents, being agents the protagonists available and present at the meeting in the act of improvisation and, performative as the potency and strength of the action per se of the one who proposes it. An action that unfolds organically into others from what belongs to one and to the collective.

Often, the term "presence" explored in the performing arts bears out the idea of nets as intertwining and weaving, in which performative agents place themselves in "contagion zones"[9] so that sharing takes place between audience and artists.

Ferracini [14] in his article *A presença não é um atributo do ator*, expands the condition of presence beyond the corporal work of the actor, dancer, musician and performer, suggesting that the event of the scene is amalgamated by the agents in its complexity, and proposes:

> The actor, as a poet of the action, should seek to build and rebuild his actions together WITH the public space and not accomplish something FOR a public space. Never a transcendent body, nor essentialist, nor solipsistic, nor endogenous

---

[9] Term created by Ferracini inserted in his publications.

but a body traversed by forces that are territorialized in the between reality/fiction, interpretation/representation dualisms. [14].[10]

We understand, therefore, that the presence effect is constituted from existences given by encounters, failures to meet and interactions.

Expanding into the ground of virtuality, Joris Weijdom [15], in his book *Mixed Reality and the Theater of the Future Fresh Perspectives on Arts and New Technologies*, presents reflections on the terms presence and reality, recognizing and including other environments in the theatrical sphere and presential arts. The author considers mixed reality (MR) as a scale on the 'continuum of virtuality'[11],that is, this scale varies '[...] from the completely real environment' up to 'the completely virtual environment' where "real environment" as the physical space our bodies are situated, where sensory entering offers a mental experience of what we consider reality. [15].

The author [15] also comments that in the mixed reality environment, the terms physical and mental, or analogue and digital, are so intertwined in the experience of reality that it is difficult to make a clear distinction in a philosophical discussion between what can be considered completely real and what cannot.

We can consider that in multimodal improvisation in mixed reality, the experience of presence happens from the state of fusion between the environments, the languages and performative agents. The different environments coexist in the space of the creation and the time of realization. Sharing between environments, agents, knowledge and conditions generate impulses of action and reaction in a dialogical and spontaneous dynamic, and the measurements given and received during the path in which the experience shared takes place.

In this sense, the concept of presence in multimodal improvisation in mixed environment is expressed in this indeterminate continuum and which shows itself as the primary element of improvisation, where the confidence of readiness to cast dynamics of response creates instant collaboration that resonates intimately between agents becoming expanded presence.

Improvising in mixed reality is composing shared presences and exchanging materialities arising from analog (real) and virtual environments at the time of action, interconnecting realities in a process of perceptive and sensory expansion. The individualized and shared content of each performative agent body is part of a single flow, merging the perceptions of the collective and the environments, creating a reality of the action in the constitution of a whole of the work.

---

[10] Original text in portuguese: O ator como poeta da ação, deveria buscar construir e reconstruir suas ações junto COM o público espaço e não realizar algo PARA um público espaço. Nunca um corpo transcendente, nem essencialista, nem solipsista, nem endógeno mas um corpo atravessado por forças que estão territorializadas nos entremeios dos dualismos realidade/ficção, interpretação/representação. [Ferracini, 2014: 235].

[11] This concept developed by Paul Milgram and Fumio Kishino in 1994.

3   Reflective Synthesis

The partnership between researchers (dance and music) from the Institute of Arts at the State University of Campinas BR, promoted impulses to investigate the field of multimodality in the interaction between dance, music and cyberspace with digital technologies.

In an updated perspective of art inserted in contemporaneity, the researchers point out, above all, the emergency of investigating collaborative and multidisciplinary artistic processes and practices. Both recognize art as a "weave", weaving in an integrated way its expression as a manifestation of the life embodied in its time. In this sense, art in the digital environment is a fertile ground to be explored in interactions with other areas of knowledge. Experience and practice are placed at the center of research, emphasizing the experimental and existential character of art. In this sense, the article discusses observed aspects of the interactive experiences between dance, sound, music and technology. One of the points of argument is in the practice of the interaction between the languages of the arts and the cyberspace, as a poetic interaction from which new methodological paths and expressive of the dancing body potentially emerge, echoing expanded movements in sound in a game between realities.

Through interactive practices there is coexistence between the communication modes in mixed environment so that the net can happen, requiring a connecting and extended body that is connected to devices and technological tools using a new body perception and movement in a creative state, unfolding and performing beyond its biomechanical limits.

In the view of the researchers, the methodology developed in the works Jardim das Cartas and "De uma Margem a Outra" is inserted in the field of complexity. Dynamic and emerging systems, self-organization and nets are the paths covered in those works. In each process, a new relational map is built or woven. The route becomes apparent (while presence) in events and poetic moments showing the importance of procedural uniqueness and articulation and interactive dynamics and dialogue between all agents. The work becomes potentially apparent in "presences" by the different states of action and levels of convergent and divergent interaction established by the agents.

Each map is constituted as a unique and complex living organism, creating its own organizational framework where interactions generate possible knots and fissures that unfold in apparent presences as condensed realities. Sound, movement and images intertwine into realities. Dance in mixed reality merges the body into virtuality and becomes "presence".

In multimodal improvisation in mixed reality, agents share space and time in action; share information and exchange movements in sound or sound in movement; in the arena of improvisation, agents experience predictable and unpredictable situations and learn the game of generosity, availability and attention to deal with the risk.

The body that dances in the interaction with the sound in mixed reality repositions itself all the time in divergent and convergent spaces searching for the sound body. Improvisation as an autonomous modality of creation and a path to the life of inte actions is, itself, the place of potency and the ultimate result.

References

1. Leman, M., et al.: What is embodied music interaction? In: The Routledge Companion to Embodied Music Interaction, pp. 1–10. Taylor & Francis Ltd. (2019)
2. Le Breton, D.: Antropologia do corpo. 4a. ed. Trad. de Fábio do Santos Creder Lopes. Petrópolis, Rio de Janeiro: Vozes (2016)
3. Nhur, A.: Do Movimento ao Som, Do Som ao Movimento: relações bioculturais entre dança e música. Rev. Bras. Estud. Presença, Porto Alegre **10**(4), e100069 (2020)
4. Leman, M.: Embodied Music Cognition and Mediation Technology. The MIT Press, Cambridge (2008)
5. Mabry, S.: Exploring Twentieth-Century Vocal Music: A Practical Guide to Innovations in Performance and Repertoire. Oxford University Press, Oxford (2002)
6. Padovani, J.H, Ferraz, S.: Proto-história, Evolução e Situação Atual das Técnicas Estendidas na Criação Musical e na Performance. Revista Música Hodie **11**(2) (2012)
7. Carvalho, R.: Técnicas estendidas para a voz: a vocalidade contemporânea nas obras de Cage, Berio, Ligeti e Schoenberg. Revista Vórtex, Curitiba, vol. 6, no. 1 (2018)
8. Sogabe, M.: Instalações interativas mediadas pela tecnologia digital: análise e produção. In: SCIArts. Ano 8 Nº 18 Metacampo, Itaú Cultural (2010)
9. Manzolli, J., Andraus, M.B.M.: Jardim das Cartas a presence ecology a multimodal and dynamic flow in an installation. In: ARTECH 2021: 10th International Conference on Digital and Interactive Arts, Aveiro, Portugal, Portugal, October (2021)
10. Manzolli, J.: Multimodal generative installations and the creation of new Art form based on interactivity narratives. In: Proceedings of the Generative Art International Conference, Veneza (2017)
11. Harispe, L.A.M.: A improvisação-dança nas coordenadas do composicional Dissertação De Mestrado Em Artes Cênicas. UFBA, BA (2014)
12. Lévy, P.: O que é o virtual? vol. 34. São Paulo (1996)
13. Ferracini, R.: A presença não é um atributo do ator. In: Linguagem, Sociedade, Políticas.1ed. Campinas e Pouso Alegre: RG e Univás. vol.1, pp. 227–237 (2014)
14. Weijdom, J.: Mixed reality and the theatre of the future fresh perspectives on arts and new technologies. Published by IETM - International Network for Contemporary Performing Arts, Brussels In partnership with HKU - University of the Arts Utrecht Original edition: March (2017)
15. Chenball, R., Kohn, T., Stevens, C.: Sounding Out Japan A Sensory Ethnographic Tour, 1st edn. Routledge, London (2020)
16. Dell'anna, A., Leman, M., Berti, A.: Musical interaction reveals music as embodied language. Front. Neurosci. **15**, 667838 (2021)
17. Leman, M.: Musical Gestures and Embodied Cognition. Actes des Journées d'Informatique Musicale (JIM 2012), Mons, Belgique, 9–11 mai (2012)

Immersive Dramaturgy. Aesthetic Dance Experiences Embodied Through Virtual Reality

Ivani Santana[(✉)] [ID]

Federal University of Rio de Janeiro, Rio de Janeiro, Brazil
ivanisantana@eefd.ufrj.br

Abstract. This study aims to explore the potential connections between dance and extended realities. This interdisciplinary approach is based on the concept of dance mediated by technological intervention. In other words, technology isn't merely used for illustration; it serves as a pivotal catalyst for physicality and aesthetic outcomes. Within this research, we're considering both virtual and augmented reality, in addition to the metaverse and imagery generated through 360-degree cameras. Understanding how these systems and tools can enhance technological mediation in dance creation is intriguing. Technological mediation can introduce novel bodily triggers and extraordinary perceptions for both dancers and the audience, aligning with the objectives of the artistic proposal. The experiences with video conferencing during the pandemic played a critical role in shaping what I've termed "immersion dramaturgy." Employing expanded realities has been essential in delving deeper into this form of narrative. The text highlights and deliberates upon the following artistic endeavors: "Itaara" (Hub, Mozilla), "ECOS" (Virtual Reality), and "Em_Corpa" (360-degree video, Augmented Reality).

Keywords: Expanded Realities · Virtual Reality · Immersion Dramaturgy · Dance through Technological Mediation

1 Introduction: The Dance Mediated by Digital Technologies

The subject of this article revolves around a dance performance conducted within the realm of virtual reality, while also encompassing an exploration of other forms of expanded reality, including augmented reality. Furthermore, the article delves into investigations involving the metaverse and 360-degree video studies.

However, it's essential to clarify that our interest doesn't lie in the functional aspect of digital devices, but rather in what we refer to as "technological mediation" [17–20]. The research and creative process pursue aesthetic objectives rooted in bodily and perceptual discoveries that emerge through this mediation between the agents of action (dancers and/or the audience) and digital technologies. In this introduction, I will present the concept of "technological mediation in dance" [20] to provide a lens for understanding the studies and work discussed in this text.

In Sect. 2, I delve into dance projects that I consider to be situated within the spectrum of expanded realities. This encompasses a journey from telematic dance [9], through

A. L. Brooks (Ed.): ArtsIT 2023, LNICST 564, pp. 73–90, 2024.
https://doi.org/10.1007/978-3-031-55319-6_6

experiences in virtual realms like Second Life, culminating in current projects called expanded realities: augmented, virtual and mixed reality. Within this reflection, I also incorporate the metaverse and the realm of 360-degree video.

In Sect. 3, I will showcase three works that I've created to explore the immersive dimensions of dance: 1) Itaara was conceived on the Hub platform (Mozilla); 2) ECOS (ECHOES), a project seamlessly intertwining dance improvisation and virtual reality; 3) "Em_Corpa" is a project that aims to develop a specific language for images generated by a 360-degree camera, considering it distinct from the pre-existing audiovisual language coming from the two-dimensional flat camera (conventional 2D camera). This provides a technical foundation for the development of 360-degree video dramaturgy itself. Through an exploration of this three-dimensional visual language, we've brought to life the interactive video and audio installation known as Em_Corpa. This project is currently in the developmental stage, presenting itself as a performative installation that incorporates images crafted through 360-degree video and augmented reality.

1.1 Dance Improvisation and Digital Technologies

From my standpoint, the utmost significance for dance in incorporating digital technology lies in the potential to unearth fresh corporeal stimuli. My focus extends beyond mere scenography or visual aspects [19]. The goal is to establish a dialogic realm where technology and the body engage in an exchange of information, facilitating the emergence of aesthetics and dramaturgy within the composition [1, 15].

Hence, within this perspective, the dance crafted through what I regard as technological mediation strives and validate this dialogic connection; and illustrate that this integration reshapes the dancer's perception, and possibly even that of the audience. Consequently, actions ensue as a result of the stimulus furnished by technological mediation. As I wrote in my book Dança na Cultura Digital [20]:

> The body of dance and technology travels in this complex melting pot of culture in permanent imbalance and transformation. Thus, dance with technological mediation should not be considered a stylistic innovation of a dance that uses new media in an indiscriminate and naive way, in the form of facilitating or decorative tools. Technologically mediated dance is an artistic manifestation that emerged from an "irremediably random" world such as that described by Ilya Prigogine, which allows us to understand the environment-individual relationship as a relationship of mutual implications. This implication consolidates the presence of the computer in everyday life and, therefore, changes the body that deals with it throughout this interaction. Therefore, the connective specificity implicit in it must not be lost, at the risk of trivializing what distinguishes it [20, p.33].

Given this conceptualization of a "live" system as one characterized by reactivity and dialogical engagement, dance improvisation appears to be better suited for this dynamic than a predetermined choreography. To sustain an ongoing dialogue, both technology and the dancers necessitate consistent interaction. This article will not discuss or reflect on dance improvisation itself. The interest here is to highlight that all the dance works presented as case studies in this text were created through dance improvisation. Reflections on improvisation will be presented in the sections of each artistic work: Itaara, ECOS and Em_Corpa.

Hence, these projects are oriented towards technological mediation in dance, and it is from this conceptual framework that I elaborate upon the projects I introduce in the following sections.

2 Expanded Realities in the Dance

Certain scholars perceive virtual reality as an illusion [6], while others contend that it delivers an experience akin to those encountered outside the virtual environment [21]. A third perspective posits that this binary categorization is, in fact, [10, 12]. I align with the latter stance, in which scholars argue that "(...) VREs can be analogous to non-virtual experiences in an important sense. Therefore, they are not illusory—yet, they fail to have the same *meaning* as (non-virtual) cognitive states" [16].

It is within this conceptual framework that we aim to contemplate dance within expanded reality contexts, primarily focusing on virtual reality, which we deem the most immersive platform. This choice stems from our intention to avoid immersing users in an illusory state where they lose awareness of their present physical embodiment and become confined to a primarily cognitive state. Similarly, we reject the notion of an identical corporeal comprehension between the virtual experience and that within the physical environment. We posit the existence of various states of presence, each implying distinct experiential dimensions [15].

In the realm of dance, this stance holds immense significance, as we aspire for this domain to serve as a realm of corporeal revelations rather than merely facilitating contemplative, metaphorical, or cerebral encounters. The veracity of this assertion will become more apparent through the forthcoming examples.

Let us adopt the metaphor of immersion, as articulated by Oliver Grau [6], yet with a keen focus on perceiving virtual reality. The expanded realities, especially virtual reality, constitute an "as if" experience, as stated by Rolla et al. [16], and it is more appropriate to use the term "allusion", therefore, distinct from understanding as an illusion.

The creative processes I have been engrossed in, facilitated by technological mediation, have consistently pursued a distinct level of immersion. Whether in stage productions or even in the realm of telematic dance, extending to contemporary endeavors within virtual reality, the overarching goal has been to reexamine, introspect, and delve into corporeality with regard to its weight, gravitational pull, dimensionality, tactile perception, olfactory stimuli, and more. This endeavor thereby evokes novel or unorthodox sensations and perceptions.

Consequently, when confronted with these alternative perspectives on the dance experience and the very act of dancing, I gradually discerned the emergence of what I term "immersion dramaturgy." I introduced this concept during the pandemic, a period marked by our profound engagement with remote interactions. Although not the primary emphasis of this article, it remains imperative to assert that these novel dance formats necessitate a departure from conventional narrative creation. The ensuing examples will elucidate several pathways in this direction.

During the pandemic, we explored the different possibilities for using videoconferencing rooms to hold artistic presentations. The public was isolated in their homes and, in general, watched the presentation alone, choosing how to participate in that meeting.

We started studying ways to encourage audience participation, whether through the camera image or through the sounds you could send through the microphone. In addition to creating a dramaturgy that encouraged the public to participate in the work's own narrative, I realized the power of using several meeting rooms simultaneously. In this sense, one of my proposals was important for the creation of the work "[In]Submersas" (2021), which featured several sensory rooms, each created by an artist from the Mulheres da Improvisação (Women of Improvisation) collective. The audience was divided into groups, which were sent to another room every 10 minutes. The objective was to promote a network of different sensations, as each artist promoted a different immersive environment, for example: one emphasized the sound issue, another was visual, another was more performative, and so on. One of the artists went through each room proposing a dance improvisation with the elements that were there. The term immersion was then used to demonstrate that dramaturgy is constructed by sensory layers in virtual environments, in which the audience is immersed. Following the metaphor of surfing the internet, we believe that diving is a verb that defines the action we propose to the public.

In another project, called Casa Parangolé, the public was encouraged to participate from the beginning, first with simple things, encouraging, but not forcing, everyone to say their name. This stimulus came with the phrase "Marielle, present!" a reference to a councilor murdered in Brazil that gained great repercussion. The audience then started saying her name followed by "present". After that, other stimuli were given, and, at a certain point, the public was thrown into rooms with different performances. When placed back in the main room, they were invited to report what they had seen in their rooms. This resulted in a dialogue between everyone present, who shared what they had witnessed during this immersion in another environment. At the end, everyone was invited to join in a dance using their "parangolés", a reference to the famous wearable work by Brazilian artist Helio Oiticica.

These experiences deepened my studies for more than 10 years in the telematics area. My current investigation into expanded realities is interested in developing "immersive dramaturgy", enabling the subject to immerse themselves in the paths that the artwork can offer them.

2.1 Telematic Dance

In a dedicated edition of the MAPA D2 Journal, Map and Arts Program in Dance (and Performance) Digital, 2015, volume 2, (www.mapad2.ufba.br) [17], I curated a compilation of articles that delved into the realm of telematic dance projects, representing over a decade of dedicated involvement. This issue includes contributions from both myself and the collaborators with whom I've engaged in this extensive journey within the domain of telematic dance.

My introduction to telematic dance occurred during a phase of my doctoral studies at Ohio State University under the guidance of Johannes Birringer, spanning from 2001 to 2002. During my time in the United States, I was involved in sessions that entailed interactions among five universities. This platform was leveraged to showcase the respective endeavors of each institution through internet-based presentations. Within these virtual gatherings, each group shared their choreographic creations with their remote colleagues.

Upon returning to Brazil, the opportunity to stage a telematic dance production only materialized in 2005, when we inaugurated the Brazilian academic high-speed internet network called Rede Ipê, an infrastructure comparable to an Internet2 connection. It was during this juncture that I received an invitation to conceive a performance, which I subsequently named "VERSUS" (2005).

In my inaugural telematic venture, VERSUS (Fig. 1), the central emphasis lay on establishing connections among geographically distant dancers. My intention went beyond mere passive observation. The platform for my presentations underwent a transformation, becoming akin to what could be referred to as a "hole in space" (and time), a concept that shares affinities with the title coined by Sherry Rabinowits and Kit Galloway for their 1980 creation facilitated by satellite.

Fig. 1. VERSUS (2005). Telematic Dance between 3 Brazilian cities: Salvador, Brazilia and João Pessoa.

Over the course of more than a decade engrossed in telematic dance creation, numerous concepts were explored to evoke a sense of connection and immerse dancers within that expansive and remote realm of space and time. Projects were undertaken to accentuate image manipulation; we devised an approach to layer live and pre-recorded footage, enabling the expansion of the stage's dimensions; we incorporated intentional data delays as an aesthetic facet to accentuate and delve into temporal considerations; novel "bodies" emerged through the amalgamation of images featuring dancers in distant dialogue; and so forth. In retrospect, I discern that during that period, immersion dramaturgy was already in the process of inception, as innovative methodological strategies were being devised to facilitate the evolution of these endeavors. As I explained in my article "Moist art as telematic dance: connecting wet and dry bodies "[The telematic dance] This new art configuration promotes different sensory-motor experiences compared with a stage-based dance environment. The networked field of telematic dance is one way in which to render reality fluid in a moist context." [19, p.187].

2.2 Second Life

Launched as a platform for virtual reality by Linden Labs in the year 2003, Second Life currently has more than one million active users who frequently interact with the simulated worlds that were developed within it.

It is a virtual and three-dimensional environment that simulates, in some aspects, the real and social life of human beings. Depending on the type of use, it can be seen as a game, a mere simulator, virtual commerce or a social network. Second Life (www.secondlife.com) allows the environment to be developed by its "residents", who adopt an avatar to inhabit this world, using it as a platform for communication, social life, and creative activities. Residents of this "second life", through their avatars, can chat with each other online, socialize, visit exhibition, and perform a multitude of "real life" rituals. The platform also allows the construction of all types of 3D objects, programming necessary tools, and so on, and all productions are listed in an inventory, a section of the interface where all the belongings are listed. The modeling of objects and avatars is based on polygons covered in textures, allowing the images to become more real.

Dance practitioners have also explored the potential of the platform as well as the metaverse today. I found it essential to incorporate this information in the text; however, this article does not aim to explore Second Life but to bring into discussion the possibilities of the metaverse in dance.

2.3 Expanded Realities Examples in Dance: My First Experiences

My inaugural encounter with augmented reality materialized through the artistic endeavor "Gretas do Tempo" (Cracks of Time) (2014) which I specifically crafted for Balé Teatro Castro Alves, the official company of the city of Salvador (www.gretasdotempo.com.br). This project encompasses three distinct artistic creations: a collection of screendance pieces titled "Memories of a memory", an interactive installation - "Memories in time" -, and a soundwalk for the dance - "Memories in space".

To access the installation situated within the historic edifice known as Palácio Rio Branco, the audience traversed a chamber where the screendances were projected onto a cube. Within this initial space, stands adorned with mirrors showcased augmented reality markers. Attendees could employ a tablet to engage with these markers, subsequently activating additional screendance components. The mirrors fostered the conception of layers, echoing the stratums we had previously explored in telematics. Furthermore, they served as a means to engross the audience within the visual poetics of the environment.

Here again, I realize that this may be a feature of what I currently consider immersive dramaturgy. The three works by Gretas do Tempo allowed the public to appreciate the artistic theme through different sensory experiences. In the case of augmented reality in Gretas do Tempo, the public accessed the image through markers fixed to a mirror. With this, the public watched the video accessed at the same time that their own image was also reflected in the mirror, which also reflected the screendances that were behind them at the entrance of the installation through which they crossed to enter the Palace Rio Branco.

In these examples, technological mediation was important for the creation of each part of the work. The process of recording the videos required the dancers to dialogue with the camera and, in addition, to be aware that these videos would be used in this

Fig. 2. Gretas do Tempo - "Memories of a memory" (2014). Augmented Reality installation.

entrance portal or in images that would be reflected in the mirror. In this way, the process is completely involved with the technological devices used. The aesthetics of the work, the movement of the dancers, and the recording of images need to take into account this proposed articulation between the media (see Fig. 2).

In 2022, I explored augmented reality in my production titled As Histórias de @eva-mariageni (The Stories of @evamariageni) (http://evamariageni.nicepage.io/). Upon entering the theater, the audience was met with three dancers positioned within the foyer. Attendees were invited to interact with images associated with markers and QR codes adorning the dancers' arms, hands, and bodies. These dancers, blindfolded and positioned atop an 80-cm diameter circular balance platform, sustained a continuous oscillation. The visuals and avatars that materialized through the spectators' tablets and smartphones stood in sharp contrast to the (almost) immobile stance of the dancers' bodies. This juxtaposition conveyed the notion that augmented reality held the potential to unveil each individual's memories and contemplations. These are other ways of understanding immersive dramaturgy that I am proposing based on this artistic research.

3   Three Experiences for an Immersive Dance

The artistic trajectory delineated above, coupled with the comprehension of the concept of technological mediation in dance, has been pivotal in furnishing a foundation for our contemporary proposition within the realm of expanded realities. This encompasses our exploration of the Hub platform (Mozilla) and 360-degree videos as well. These creation processes are important for the development of immersion dramaturgy based on technological mediation in dance. It is worth mentioning that the relationship with the camera (whether virtual reality or 360-degree video) was created from a dance improvisation, which, in this case, had a three-dimensional and immersive image as a partner. This specific configuration carries aspects that demand other sensorimotor contributions and stimulate other aesthetic possibilities. The following pieces will be presented: Itaara made on the Mozilla Hubs metaverse platform; ECOS, a virtual reality project created with Unity; and Em_Corpa, a performance installation that integrates 360-degree videos and interactive sound.

3.1 Itaara. A Dance Experience on Mozilla's Hubs Platform

Itaara played a fundamental role in shaping our considerations surrounding an immersive, three-dimensional, and interactive environment akin to virtual reality, despite its conception within a web platform. This endeavor unfolded concurrently with the evolution of the virtual reality venture ECOS, a topic we will delve into in the ensuing section. Given that the level of intricacy on the Hub platform is comparatively more approachable than the development of a virtual reality project constructed through Unity, this experience assumed a significant role within the project. It enabled us to contemplate the prospects for audience engagement with the work, in addition to devising the most effective means of integrating dance within that contextual framework.

The conceptualization and development of Itaara were undertaken by the team at Technological Poetic Research Group: corpoaudiovisual throughout the period from 2019 to 2021. Derived from the Tupi-Guarani language, Itaara translates to "high stone" or "stone altar." The selection of this title was influenced by my time spent during the pandemic at Itacimirim Beach in Bahia, Brazil. Situated amidst stone formations that naturally shape pools, this coastal haven conferred a sense of privilege, despite the ongoing weight of isolation and the separation from family members. Additionally, this locale served as an evocative backdrop for filming, utilizing both conventional flat camera (2D) and immersive 360-degree (3D) video captures, which were extensively employed in the creation of various environments designed for the Hubs platform.

The project was guided by myself, Ivani Santana, and had Renan Felipe Bolcont de Menezes and Letícia Mayne in the development of the virtual environments, while Camila Soares was responsible for the screendances with me.

Our initial inquiry revolved around a fundamental question: How could dance be perceived and not merely observed within this environment? Our intention diverged from creating a conventional gallery of framed images. Instead, we pondered how to engender sensory experiences for users by leveraging the potential afforded by the Hubs platform, even while navigating its inherent limitations. Given the ubiquitous influence of the mediation concept across all project phases, a paramount consideration was to embark on an exploration of the system, intertwined with the established knowledge pool in the realm of media art. Furthermore, this exploration intersected with the realm of dance enriched by technological mediation.

For our team, the paramount attribute under scrutiny centered on the exploration of a three-dimensional environment, heightened by the sophisticated integration of spatialized sound. This foundation prompted our pursuit of crafting distinct experiential trajectories for users throughout their engagement. Within the inaugural chamber, users were presented with detailed directives pertaining to the navigation of portals that grant entry into subsequent chambers, coupled with comprehensive instructions elucidating how to navigate the environment through the utilization of either the mouse or keyboard arrow inputs.

This initial room (Fig. 3), for instance, comprised an endless expanse of black space, offering users a stone pathway along which they traverse images of the "sphinx woman" – a figure that forewarns about the realm of Itaara and beckons them to embrace the experience of embodying an Itaara being themselves.

Fig. 3. Itaara (2021). Hubs, Mozilla. The entrance. Technological Poetics research group: corpoaudiovisual. https://hubs.mozilla.com/ArbFzkU/slight-natural-plane.

Amidst the vast expanse of the initial space, users experienced a sense of liberation and acquired the ability to navigate amongst the evocative depictions of the sphinx woman. Spatialized auditory cues further facilitated an enhanced grasp of the three-dimensional environment. As users approached the auditory objects, the sound intensified, diminishing in volume as they distanced themselves from the source.

Fig. 4. Itaara (2021). Hubs, Mozilla. The Chaos room. Technological Poetics research group: corpoaudiovisual. https://hubs.mozilla.com/mkw436M/caos.

Conversely, the second room (Fig. 4) incarcerates the user within a tumultuous milieu characterized by boisterous and forceful auditory stimuli. The swift and vigorous alternation of vivid images projected from all facets of this cubical space intensifies the immersive experience. The portal serves as an egress for the user, guiding them towards another sensory realm.

Fig. 5. Itaara (2021). Hubs, Mozilla. The contemplation room. 360-degree image. Technological Poetics research group: corpoaudiovisual. https://hubs.mozilla.com/apfDPxM/mundo.

The entrance of this hallucinatory expanse guides the way to access a new environment (Fig. 5), now dedicated to a moment of serene contemplation. Within this setting, a 360-degree photograph invites users to engage in unhurried contemplation, encouraging them to leisurely survey the entire encompassing space. This perspective assists users in identifying subsequent portals that will lead them into further sensory realms.

In this manner, users are incited to navigate the spatial domains, thereby revealing an assortment of distinct sensorial landscapes. Within one such environment, suspended stones come into view, and upon crossing them, they unveil auditory messages. In an alternate locale, users are invited to explore a labyrinthine configuration of stones adorned with 2D depictions of the performer (Fig. 6). A quintessential characteristic of labyrinths, specifically the sensation of aimless wandering intertwined with the pursuit of an exit, is unmistakably established. Furthermore, an environment boasting a 360-degree video format is presented, summoning active contemplation akin to its photographic counterpart. Users are enjoined not solely to search for the portal, but also to manipulate the image's proportions, thus transmuting it into a spherical representation (Fig. 5). Lastly, an enclave is designated wherein users can activate both their camera and microphone (Fig. 7), evoking a culminating conversational circle. In facilitating the presence of other users, this space engenders opportunities for interpersonal exchange and discourse, mediated through the technological device.

These dynamics were integral to the dramaturgy of Itaara, designed to provide users with stimuli connected to movement, distinct spatial sensations (such as vastness, confinement, and labyrinthine structures), and diverse temporal experiences (including contemplating photographs or videos, and navigating intricate paths). Itaara stands as an example of what I refer to as "immersion dramaturgy," a concept that invites and encourages the audience to engage with a range of sensations by immersing themselves in various virtual environments. In this regard, the primary aim of Itaara within the Hubs platform—a virtual web environment—was to bring the essence of dance into the user's encounter with the platform.

In Itaara, the audience's experience embodies an 'as if' sensation, rooted in the illusion of vastness, entrapment, or labyrinthine exploration, contingent upon the specific room they occupy. When viewed on a computer via the web, the sense of immersion tends to be more subdued compared to physically inhabiting a space with these attributes, or even within a virtual reality environment that elicits heightened engagement of the entire sensory apparatus. Regardless of the medium, it remains an allusion, as posited within this text.

Contemplating a narrative rooted in sensations, poetic resonance, and embodied connections, rather than adhering to a framework centered on gaming and competition, emerged as a propitious trajectory for the development of the virtual reality endeavor known as ECOS.

Fig. 6. Itaara (2021). Hubs, Mozilla. The Maze Room. Technological Poetics research group: corpoaudiovisual. https://hubs.mozilla.com/V9i54r8/caminhante.

Fig. 7. Itaara (2021). Hubs, Mozilla. The confessional room. Technological Poetics research group: corpoaudiovisual. https://hubs.mozilla.com/HHgKPPi/eu-sou-pedra-oceano.

In the case of Itaara, we can say that the characteristics of a dance improvisation process were released to the public. At each moment, in each room, the audience is involved with the environment and, in this interrelationship, they decide which paths to take and how to move their avatar to dialogue with that environment and its elements. What matters in the improvisation process is precisely the choices we must make when faced with a stimulus. This aspect was chosen to be explored in Itaara, and it was based on this factor that we created rooms with different sensory provocations to offer the public.

This artistic research then seeks to apply knowledge from the field of dance to deepen the notions of technological mediation, immersion dramaturgy, and even improvisation itself.

3.2 ECOS. Dance and/in Virtual Reality

ECOS represents another path through which I have strived to delve into the immersive dramaturgy I have been cultivating. This project presents an alternative approach to harnessing virtual reality, transcending the domain of games and educational applications. The integration of virtual reality into the realm of contemporary dance remains relatively new, particularly at the intersection of these two domains, with only a handful of researchers and artists currently exploring this convergence. ECOS introduces a dance improvisation meeting anchored in uncomplicated gestures that arise from the interactivity evoked by the constituents of the virtual realm.

ECOS was conceptualized as an interactive realm, fostering engagement among the user, the avatar, and the virtual environment. The user's actions manifest within the environment, inducing alterations encompassing chromatic shifts, physical transformations, auditory variations, and even the avatar's embodiment. This virtual ecology engages in a symbiotic interplay with the rapport forged between the user and virtual reality, reliant upon the participant's motions for its unfolding. Our objective resides in furnishing visitors with an encounter that heightens their awareness of their own corporeal dimensions contingent upon their engagements within this digital realm. A poetic proposition that delineates how our conduct can reverberate across the environment we inhabit.

User engagement necessitates active involvement, wherein participation, perception, and interaction harmoniously intertwine with the environment. Their gestures evoke responses within the spatial dimensions, impacting facets such as color, saturation, and clarity, among others, while simultaneously influencing the avatars within the system, fostering a seamless connection with the participants. The caliber of interaction—comprising elements like intensity, duration, and scope of action—serves as the determinant of the degree of interconnectedness linking the user, the avatar, and the system. A heightened level of contact, encompassing both visual and gestural aspects, amplifies the vigor of their actions—be it through movement, locomotion, or gestures—culminating in heightened engagement and perception.

The inception of the project was grounded in an exploration of improvisational procedures within the domain of dance, viewed through the analytical frameworks of Enactivist Theory [23] and Linguistic Bodies [2]. Within this theoretical framework, we conceive the actions inherent in improvisation as outcomes engendered by the phenomenon of "participatory sense-making" [2]. Hence, improvisation transcends individualistic decisions and unfolds as a convergence of individual agency interwoven with the actions of fellow participants and the contextual fabric, collectively shaping the landscape of decision-making. This interplay of actions resonates within the system, prompting a reciprocal cascade of effects across the process and the realm of corporeality. Anchored in this theoretical backdrop, the project embarked on an investigative trajectory, seeking to dissect the intricate web of interrelations among the constituents of a given system. In the context of ECOS, these constituents encompass an array of entities, including the user, the avatar, diverse elements, entities, and the very spatial environment—each contributing to the complex network of interactions.

The project presented several challenges for its development in Virtual Reality. The first was to create a library of actions and reactions between the avatar's movements and the user's movements. The head-mounted display sensors impose a certain limitation on capturing a full body expression, so the system needs to infer the user's gestures from these three parameters (glasses and 2 hand controls) added to their movement in the action area. The displacement is considered, in principle, within the avatar's area of influence. In turn, adequate programming had to be generated to capture these movements and gestures and generate a dynamic response in our virtual interlocutor. In addition, it was sought that this interaction also influence the environment, from shapes, materials (shaders), and various parameters that affect the appearance of the environment and its elements. This interplay further extends to induce emotive responses within users, cultivating a holistic experience that encompasses affective dimensions.

ECOS comprises two discrete environments – a cave (Fig. 8) and a lake (Fig. 9) – embellished with mountains and a bridge, serving as the point of entry for users upon their ingress. Echoing the conceptual underpinning of Itaara, the fundamental idea was to propose a dyadic topography: one encapsulated and the other expansive. Distinctive responses from both the environment and avatars demarcate these realms. However, a unifying thread interlaces them: the imperative for users to engage in corporeal move-ments, particularly those involving their hands and head, in order to render all elements discernible. Within the cave precinct, for instance, an initial veneer of whiteness is punctuated by subtle contours and shadows. The user's movements engender escalated saturation, progressively unveiling all constituents and enhancing environmental per-ceptibility. Conversely, upon venturing into the lake expanse, the user is greeted by the bridge, mountains, sea, and sky. As the user navigates, mountains flourish with verdant foliage, and the realm is graced with fanciful entities like whales and airborne fish – all catalyzed by the user's movements. Across both domains, the avatar assumes a humanoid form adorned with leaves and plants in the lake, or imbued with brushstrokes akin to cave paintings in the cave – a manifestation that harmonizes the tangible and the ethereal.

Although we still wish to develop ECOS, placing other rooms and other move-ments for the avatars, it is important to state that even at this current stage, the bases of dance improvisation, technological mediation, and immersion dramaturgy are effec-tively applied in this project. The notions of participatory meaning can be perceived in the dialogue established between the user and the head-mounted display, and the avatars, and the ECOS environment itself. To reach this stage, it was necessary to phys-ically verify on the dancer's body which movements would be interesting to bring the avatar to life, which is why we consider the importance of understanding technological mediation. In the same sense, it was through improvisation processes with the avatar that we discovered the best ways to implement the avatar's computational programming in its relationship and reaction with the public. Once again, the dramaturgy of immersion is installed from the moment the audience is able to immerse themselves in the work. In the case of ECOS, this is a literal condition, as the environment, avatars and virtual beings will only appear if the user with the head-mounted display moves around the

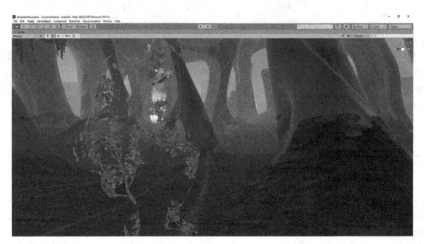

Fig. 8. ECOS (2021). Dance and Virtual Reality. The cave and the avatar as a "nature man". By Ivani Santana and Daniel Argente.

Fig. 9. ECOS (2021). Dance and Virtual Reality. The mountains. By Ivani Santana and Daniel Argente.

space and interacts with the surroundings. If there is no action, or movement from the public, nothing will happen.

Dancing with the avatar is not the same as dancing with a person in the physical world, but, following the arguments of Rolla et al. [16], it is an "as if", an allusion to improvisation. We also know that the user can move their body in a very simple way, performing movements to access the virtual world as needed by the system. However, these are not functional movements, with an end in themselves (picking up an object, firing a weapon, to give common examples in games), but moving freely through space. This is the characteristic of the improvisation that we apply in ECOS, whose objective is not specific and codified movements of any dance, but the simple possibility of acting in the space by interacting with it.

3.3 Em_Corpo: Dance on Video 360-degree

The integration of 360-degree video into our creative process commenced during the inception of the Itaara project. However, it was only in 2022 that the Technological Poetic Research Group: corpoaudiovisual embarked on a systematic exploration into the refinement of the audiovisual language inherent to 360-degree video.Although we introduced 360-degree video in the context of Itaara's creative development, the comprehensive investigation into the maturation of the audiovisual idiom associated exclusively with 360-degree video was formally carried out by the Em_Corpa creative process, which began in 2022 and will debut in November 2023 at the ArtsIT Congress, UNICAMP, Brazil.

In the realm of traditional flat camera methodologies, encompassing aspects like framing, composition, camera kinetics, and the like, a multitude of established norms exists. Paradoxically, our inquiry revealed a paucity of substantive guidance directing us towards the articulation of an audiovisual vernacular that effectively accommodates the unique attributes inherent to the 360-degree camera. In contradistinction to the prevailing inclination among certain authors and artists [1, 10, 15], who seem to endeavor to harness the viewer's perspective through the prism of established audiovisual norms, our approach takes an antithetical trajectory. We posit that the most efficacious approach ought to diverge from the endeavor of curtailing the viewer's gaze and instead, should be predicated on engendering a multitude of divergent trajectories.

In contrast to the conventional frontal engagement associated with the 2D camera, which progresses from a confined spatial frame (extreme close-up) to a broader panorama (full shot), the 360-degree camera introduces a fundamentally different structural paradigm. This configuration entails a circular interrelation, with camera perspectives revolving around the viewer's field of vision. As elucidated through the schematic representation (Fig. 10), the conventional vertical planes depicted by the 2D camera yield an alternative organizational principle. In lieu of these distinct planes, a succession of circumferential rings takes precedence, orbiting in alignment with the panoramic image captured by the 360-degree camera. As a consequence, this reconfiguration engenders a novel methodology for narrative composition, characterized by the evolving spatial dynamics inherent to the 360-degree dancing

Founded upon this comprehension, we conceived the Em_Corpa installation, wherein attendees are furnished with virtual reality headsets encapsulating 360-degree recorded imagery of the four dancers. Concurrently, the dancers themselves, each equipped with mobile devices and Bluetooth speakers, partake in the performance milieu, generating an auditory landscape interwoven with their bodily movements. The 360-degree visuals are further projected onto peripheral screens, meticulously preserving the original dual-globe configuration captured by the camera. It is imperative to note that this endeavor remains a work in progress, hence the omission of accompanying visual assets at this juncture. The installation will also incorporate augmented reality in a manner akin to what was employed in the production "As histórias de @evamariageni", utilizing markers on the dancers' bodies.

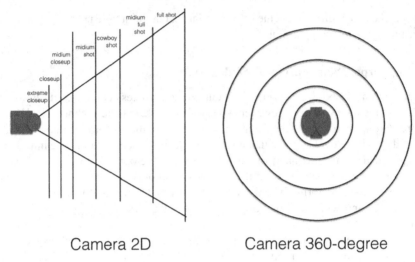

Fig. 10. Camera 2D language, and 360-degree video language.

This work is still in progress, but we already have a promising path achieved through research guided by practice. Mediation with the 360-degree camera has demonstrated aspects that are of great importance for the dramaturgy of immersion, such as opening the audience up to choose their narrative and delve into possibilities found at the moment of enjoyment. Once again, improvisation was an important process to discover this dance involved with a global image that exposes 360 degrees of vision. Likewise, the public is offered to dance in this installation, sometimes immersed in the head-mounted display, sometimes participating in the environment and interacting with the dancers. The narrative was created thinking about the transition from virtual to real space, integrating the public in these two environments.

4   Conclusion

While ECOS offers an immersive experience that encourages user movement and improvisation, it's essential to recognize the fundamental differences between the real-world experience and the virtual one. Although ECOS aligns itself with dance improvisation, primarily from a first-person perspective, there remains a noticeable distinction in the experiences provided by these two separate realms. In this complex context, we introduce the concept of 'allusion' as a compelling idea that captures certain aspects of the physical experience within a parallel virtual environment. This framework enables individuals, including those less inclined to dance, to embrace the opportunity to engage in choreographic interactions alongside digital avatars. Built upon the foundation of 'as-if,' this proposition fosters an environment where people can navigate the world of dance with increased confidence and freedom. We argue that this experience is valid due to its evocative 'as-if' nature, thereby transforming the virtual space into an empowering sanctuary for those who might otherwise avoid medium in the physical world.

The term "allusion" [16] represents a powerful idea that serves as a bridge between two distinct realms: the physical and the virtual. In this context, "kinetic encounters" refer to experiences involving movement and physical engagement, such as dance. The core idea is that "allusion" acts as a conduit, allowing people to share and participate in these kinetic experiences despite the barriers posed by the tangible (real-world) and the virtual (digital) environments.

This concept becomes particularly valuable because it offers an opportunity for a wide range of individuals to overcome their reservations or hesitations and become part of the captivating world of dance. In other words, it empowers people who might not feel confident or comfortable dancing in the physical world to engage with the art form within the virtual realm.

By using the term "allusion," this concept acknowledges that the experience in the virtual space is not a perfect replication of physical dance but rather an immersive and evocative representation that allows individuals to feel as if they are participating in the dance world. This can be incredibly liberating and inclusive, as it encourages people to explore and enjoy dance in a virtual environment, breaking down barriers and making the art form accessible to a broader audience.

The increasing diversity of experiences emerging from various corners of the world, blending the realms of dance, virtual reality, augmented reality, the metaverse, and 360-degree video, is akin to a rising wave pushing dance forward through the channel of technological mediation. Within this intricate convergence, virtual reality emerges as a prominent driving force behind this transformative journey. Our eager anticipation lies in the hope that these interconnected endeavors, spanning academia and artistic creation, will produce substantial insights, enhancing the multifaceted aspects of exploration and involvement nested within this dynamic intersection.

References

1. Sheikh, A., Brown, A., Watson, Z., Evans, M.: Directing attention in 360-degree video **29**(9) (2016). https://doi.org/10.1049/ibc.2016.0029
2. Di Paolo, E.A., Cuffari, E.C., De Jaegher, H.: Linguistic Bodies: The Continuity Between Life and Language. MIT Press, Cambridge (2018)
3. Echeverría, J.: Teletecnologías, espacios de interacción y valores. Teorema,Revista Internacional de Filosofia, n.2, p.5 (1998). http://www.oei.es/salactsi/teorema01.htm. Accessed 10 set 2023
4. Gallagher, S.: How the Body Shapes the Mind. Oxford University Press, Oxford (2005)
5. Grabarczyk, P., Pokropski, M.: Perception of affordances and experience of presence in virtual reality. AVANT. J. Philos.-Interdisc. Vanguard **VII**(2), 25–44 (2016). https://doi.org/10. 26913/70202016.0112.0002
6. Grau, O.: Virtual Art: From Illusion to Immersion. MIT Press, Cambridge (2003)
7. Hovhannisyan, G., Henson, A., Sood, S.: Enacting virtual reality: the philosophy and cognitive science of optimal virtual experience, pp. 225–255 (2019). https://doi.org/10.1007/978-3-030-22419-6_17
8. Kvisgaard, A.: Frames to zones: applying mise-en-scène techniques in cinematic virtual reality, pp. 1–5 (2019). https://doi.org/10.1109/WEVR.2019.8809592
9. MAPA D2 Journal, Map and Arts Program in Dance (and performance) Digital, 2015, volume 2. https://periodicos.ufba.br/index.php/mapad2/issue/view/1184. Accessed 10 Aug 2023

10. McMahan, A.: Immersion, engagement, and presence: a method for analyzing 3-D video games. In: Wolf, M.J.P., Perron, B. (eds.) The Video Game Theory Reader, pp. 67–86. Routledge (2003)
11. Murray, J.H.: Hamlet on the Holodeck. The Free Press, Los Angeles (2016)
12. Nilsson, N.C., Nordahl, R., Serafin, S.: Immersion revisited: a review of existing definitions of immersion and their relation to different theories of presence. Hum. Technol. **12**(2), 108–134 (2016). https://doi.org/10.17011/ht/urn.201611174652
13. Nöe, A.: Action in Perception. MIT Press, Cambridge (2006)
14. Nöe, A.: Varieties of Presence. Harvard University Press, Cambridge (2015)
15. Pope, V.C., Dawes, R., Schweiger, F., Sheikh, A.: The geometry of storytelling: theatrical use of space for 360-degree videos and virtual reality, pp. 4468–4478 (2017). https://doi.org/10. 1145/3025453.3025581
16. Rolla, G., Vasconcelos, G., Figueiredo, N.: Virtual reality, embodiment, and allusion: an ecological-enactive approach. Philos. Technol. **35**(4), 1–30 (2022)
17. Santana, I.: First Telematics Experiences of the Grupo de Pesquisa Poéticas Tecnológicas: Corpoaudiovisual. Electronic Magazine MAPA D2 - Digital Dance (and Performance) Arts Program and Map, Ivani Santana (Org) Salvador: PPGAC 2(2), 298–314 (2015). https://per iodicos.ufba.br/index.php/mapad2/article/view/14932/10231
18. Santana, I.: Novas configurações da Dança em processos distribuídos das Redes. Plataforma Eletrônica Internacional Xanela Comunidad Tecno Escenica (2013). http://www.xanela-red e.net. Accessed 10 set 2023
19. Santana, I.: Moist art as telematic dance: Connecting wet and dry bodies. Technoetic Arts 13 (2015). https://doi.org/10.1386/tear.13.1-2.187_1
20. Santana, I.: Dança na Cultura Digital. Salvador: EDUFBA (2006). https://doi.org/10.7476/ 9788523209056
21. Slater, M.: Place illusion and plausibility can lead to realistic behaviour in immersive virtual environments. Philos. Trans. R. Soc. B: Biol. Sci. **364**(1535), 3549–3557 (2009). https://doi. org/10.1098/rstb.2009.0138
22. Slater, M.: Immersion and the illusion of presence in virtual reality. Br. J. Psychol. **109**(3), 431–433 (2018). https://doi.org/10.1111/bjop.12305
23. Varela, R., Thompson, E., Rosch, E.: The Embodied Mind: Cognitive Science and Human Experience. MIT Press, Cambridge (2017)

Dance and New Technologies: Different Interdisciplinary Approaches in Teaching and Practice

Isadora Alonso Faustino$^{(\boxtimes)}$ and Daniela Gatti

Corporal Arts Department, Arts Institute, Unicamp University, Campinas, Brazil
isadora.alonso.ia@gmail.com, danigati@unicamp.br

Abstract. The present article's goal is to present reflections on the integration of technologies in the field of dance, whether in a context of performance environments or teaching-learnings ones. It also shares experiences and insights about interdisciplinary projects that involve dance, music, and new technologies. Therefore, the main goal is to showcase the challenges faced and the benefits of this interdisciplinary approach, as well as the opportunities that arise from the intersection of these different fields. Supported by a brief historical context, the potential of technology's use in relation to dance is explored, both from the perspective of the researching artist and that of the educator. Finally, the article seeks to understand how to expand the accessibility of this interaction between dance, music, and technology in various contexts. The ultimate aim is to bring the research conducted within the university to broader audiences and locations.

Keywords: Dance · New Technologies · Interdisciplinarity

1 Introduction

The convergence of technology and telecommunications has had a significant impact on the relationship between people. With computers and mobile devices, distances have been shortened, and the dissemination of information has become much faster. Postman (1994) discusses the surrender of culture to technology and elaborates on the effects resulting from the integration of these new technologies:

> The users were affected by an unbridled information avalanche that presents in its diagnosis an information disorder unable to meet the needs if its user is not prepared for its production, communication, and use (Postman 1994, p. 204)[1]

---

[1] Original text: [...] Os usuários foram afetados por uma avalanche informacional desenfreada que apresenta em seu diagnóstico um distúrbio de informação incapaz de atender às necessidades caso seu usuário não esteja preparado para a sua produção, comunicação e uso. (Postman 1994, p. 204).

A. L. Brooks (Ed.): ArtsIT 2023, LNICST 564, pp. 91–106, 2024.
https://doi.org/10.1007/978-3-031-55319-6_7

Technologies, understood here as devices created to facilitate human life, will have their focus of analysis in this paper concentrated on communication, information, and audiovisual technologies. What we are experiencing today is a new relationship with technologies, with the merging of reality and virtuality, where technological innovations begin to dictate a new societal functioning. Author Alessandra Bittencourt, when discussing the effects of this technological revolution, states that "brought changes and affected interpersonal relationships, as well as the relationship between humans, between humans and machines, between humans, machines, sciences, and arts." (Bittencourt 2005, p. 3).[2]

Dance, as a significant language and artistic expression of our culture, has also undergone transformations due to its interaction with the electronic era, as our bodies carry new traits and thought structures resulting from the interaction established with this digital and computational age, which greatly impacts dance creation. Merce Cunningham(see footnote 1) revolutionized dance with creations that employed audio, video, and motion capture technologies from the mid-1960s until the 2000s, when he began incorporating computers. Body, space, sound, image, sensors, and devices were utilized in service of dance, in order to create multi-linguistic works that served as a breeding ground for numerous interdisciplinary studies in dance up to the present day. Among them, one of the most prevalent would-be video dances, dance creations made exclusively to be recorded on video and displayed on screens.

Body-related research that engages with technologies is in constant evolution, and due to our current era of heightened interaction between the real and the virtual, new demands emerge to ensure the relevance of this type of work. Therefore, it's important to grasp that these works are always subject to change and, as Cunningham (1994) notes, they are in a process that can often lead to the development of future endeavors.

> My work has always been in process. Finishing a dance, I have the idea, often weak at first, for the next one. In this way, I do not conceive each dance as an object, on the contrary, I see it as a small stop on the way (Cunningham 1994, in Vaughan 1997a, p. 276)[3]

The social isolation we experienced in the years 2020 and 2021, due to the pandemic, marked a significant milestone for the production of videodances worldwide. The cellphone, which was already a prevalent tool in the artists[4] daily lives, became integrated into many compositions and facilitated access to video editing apps, making the creation

---

[2] Original text: "Trouxeram mudanças e afetaram as relações interpessoais, assim como a relação do homem com o homem, do homem com a máquina, dos homens e das máquinas com as ciências e as artes." (Bittencourt 2005, p. 3).

[3] Original text: Meu trabalho sempre esteve em processo. Terminando uma dança, fica-me a ideia, frequentemente fraca no início, para a próxima. Dessa forma não concebo cada dança como um objeto, ao contrário, vejo-a como uma pequena parada no caminho. (Cunningham and Vaughan 1997a, p. 276).

[4] Merce Cunningham (1929–2009) began his professional career at age 20 as a soloist in the Martha Graham Dance Company. Over his long career, Cunningham welcomed new technologies, working with dance on film, computer programmed chore ography, and even motion capture technology for the piece Biped (music by Gavin Bryars) in 1999.

and study of dance even more accessible. Menicacci (2004) discusses this process of democratization of technology access and highlights its effects on dance.

> The democratization of motion capture could open new perspectives for teaching dance. There are many instruments and techniques, and they will multiply rapidly, also becoming more accessible. It is necessary to introduce them in dance education. (Menicacci 2004[5])[6]

Within this context, my journey as an artist, educator, and dance researcher comes into play. I strive to explore an interdisciplinary connection between dance, music, and new technologies in order to comprehend alternative ways of working with dance, while ensuring it retains its essence rooted in the human body. This essence doesn't rely on any technological apparatus to exist, yet it has been and continues to be influenced by this new way of interacting, understanding, and existing in the contemporary world.

This article aims to demonstrate how technology can be integrated into the teaching and practice of dance, considering its evolution and increased utilization, and contemplating the social implications and new habits that have emerged during and after this period. The theoretical foundation was built through examples of projects designed and experienced throughout the history of dance, engaging primarily with the works of Merce Cunningham and Daniela Gatti. For the analytical aspect of these projects, the dialogue is established with Alessandra Bittencourt, Ivani Santana, and Lucia Santanaella.

To achieve this, it is anticipated that the interdisciplinary relationship between technology and dance practice will demonstrate great potential. Beyond enhancing the artistic experience, it has the potential to serve as inspiration for the new generation, which was born and raised in a digital environment. Furthermore, the aim is to illustrate how this strategy can be employed and adapted to various contexts without overshadowing the physical work and bodily experience intrinsic to dance.

2 Interdisciplinary Experiences Between Dance, Music, and New Technologies

To expose the body to a technological environment means to make it sensitive to different stimuli and consequent reactions. When an artist decides to create within this context, they assume that their body won't be the highlight of that work, but instead that the work won't exist without this dialogue. In different and multiple situations, the connection between art and technology becomes the fundamental link for a work to happen.

---

[5] Citation referring to the text: "O ensino da dança face à tecnologia digital" by Armando Menicacci in Portuguese by the Consulate of France in Rio de Janeiro, available at <www.idanca.net>.

[6] Original text: A democratização da captura do movimento poderá abrir novas perspectivas para o ensino da dança. Os instrumentos e as técnicas não faltam, e vão multiplicarse rapidamente, tornandose também mais acessíveis. é necessário introduzilas no ensino da dança. (Menicacci 2004).

It was from this perspective that the interactive performance of dance and music emerged at the *Jardim das Cartas*[7] exhibition, in 2022 at the Art Gallery of the Unicamp Institute of Arts, in which the authors had the opportunity to have a performative and improvisational experience in mixed reality. This performance was in partnership with the research group *Núcleo de Dança Redes: Processos Criativos em Redes de Saberes.* In this, virtuality and reality were mixed through projections of pre-recorded videos that were influencing the dance movement of the creative interpreters. The movement, in turn, dialogued with the music, created live by the musicians Manuel Falleiros and Alexandre Zamith, the second one also being a teacher[8], who integrated the work from a distance, just following it via live transmission. The cell phone also composed the work, in the form of an app called *Movie Guitar*[9], which allowed the performers to create sound overlays with this sound generator, triggered by the movement of the device. Sound, image and movement walked towards a conceptual and constructive identification, in order to create a dissolution of boundaries between visuality and sonority (Figs. 1 and 2).

Fig. 1. Jardim das Cartas (2022). Source: Personal Archive. Picture: Diogo Angeli

Collaborative works like this work under a different organization, since to exist they depend on the harmony between all these languages that, in the context presented above,

---

[7] Jardim das Cartas was a poetic proposal conceived by Jônatas Manzolli, a professor in the music department at Unicamp, based on letters and authorial poems. From them, videos were created by the professor of the body arts department at Unicamp, Mariana Baruco, which served as material for the interactive work. This performance was open to a group of students supervised by professor of the body arts department Daniela Gatti, who also is the leader of the research group *Núcleo de Dança Redes.*

[8] Professor in the area of piano and chamber music at the Institute of Arts of the State University of Campinas – UNICAMP.

[9] The Movie Guitar application was created by Professor Jônatas Manzolli. Its operation is activated through the cell phone's gyroscope, so that when it is moved, sounds are emitted. These sounds are simple chords, which can be selected manually, in the way that best pleases the user. Therefore, when the dancer activates the application and holds the cell phone while dancing, it will emit sounds as it moves.

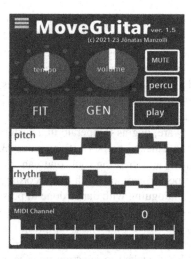

Fig. 2. App MoveGuitar. Source: Personal Archive.

do not exist individually. The dialogue between dance and technology is very significant for the work and it needs to be flexible to adaptations, since it will undergo modifications according to the type of structure of the space it will occupy. The spaces, in turn, even if they are more conventional, as is the case with Italian stages, have different physical structures according to each place, which directly impacts the development of the work. Factors such as internet connectivity, availability of power outlets or structures to support certain equipment, have a great influence on the works.

It is observed, in dances present in works such as the one mentioned above, that the relationship between the movements and the space suggests a different type of thinking, since what will dictate how the dance will succeed is precisely the disposition of the materials and tools technologies across space. The choice of movements, which in turn dialogue with the stimuli offered by technologies, will also vary according to the infrastructure. The placement of the performers in the space is often done in order to allow a more adequate dialogue with the technologies and, therefore, the choice of choreographic composition must be made in order to allow a greater capacity for adaptation from one performance to another.

In addition, all layers of technology used instigated a new perception, which influenced the way dance was perceived and performed. The movement should be emphatic in order to be able to generate sounds through the application and the positioning of the interpreters in the space should be thought out in such a way as to allow the association between movement and image. Dance was not the focus of creation, much less music or media, but the relationship between languages. This would be an interdisciplinary approach that, as elaborated by Gatti (2018), "provides a conceptual framework for working the visual, sound, bodily and spatial aspects of interaction and technological environments, and their relationships" (Gatti 2018, p. 02).[10]

Body research for hybrid performances - perfomances that mix dance and technology and use this tool to build new forms of movement - can be designed so that there is a dialogue between the body and digital technological devices. Author Ivani Santana, when discussing about this dialogue, which is established when working with dance with digital technological devices, says that "the computer, even when used as a tool by choreographic creation software, begins to operate in a system of permanent dialogue with the creator and with the dancers' bodies". (Santana 2003, p. 7)[11]. Understanding the computer as a component and creation tool, one can cite its use to allow video transmission, which can be performed both with live recordings and with pre-recorded scenes, as well as a way of dialoguing with dance done in person.

Working with dance in relation to video or audiovisual, provides the opportunity to rethink the construction of movement, either by capturing a figure from an angle that our gaze has never seen or by affecting time, that is, the dancer's rhythm. With this possibility, the viewer can be provoked to question what is actually happening in the real moment, because the live dance can dialogue with the videos, which can give the impression that they are being transmitted live, when in reality they are prerecorded recordings and vice versa.

The *Interferências*[12] concert relies on a lot of improvisation work, which is influenced by visual, digital and sound elements that are presented during the show, such as videos and live broadcasts. The use of the camera during the presentation, for example, is related to dance to form movements and experiences that would not exist, completely, without the presence of these technological devices. At one point in the performance, the dance is performed in spaces outside the theater and the movements are captured live by a video that is projected in real time to the audience. This action enables the overlapping of spaces, movements and presentation dynamics, generating an interaction between the internal and external spaces, which are the theater and the exterior, respectively, in addition to providing the public with an expansion of the perception of the performance, which is not limited to, to the spatial limits imposed by the stage.

In addition, even when it is possible to see the dance being performed in person, the video resource is used so that new details and other angles are presented to the viewer, through the mixture of real images and virtual images, which are presented in the form

---

[10] Original text: "fornece um arcabouço conceitual para trabalhar os aspectos visuais, sonoros, corporais e espaciais de interação e ambientes tecnológicos, e suas relações." (Gatti 2018, p. 02).

[11] Original text: "o computador mesmo quando empregado como ferramenta pelos software de criação coreográfica, passam a vigorar em um sistema de diálogo permanente com o criador e com os corpos dos bailarinos." (Santana 2003, p. 7).

[12] *Interferências* was a final paper of the bachelor's degree in Dance at Unicamp, supervised by Professor Daniela Gatti. It had partnerships with audiovisual researcher Guilherme Zanchetta and musician Eduardo Koji, in a collaborative construction between languages. This dance show was approved in the *12º Edital do Programa Aluno Artista* [12th Notice of the *Aluno Artista* program], which is part of student support policies in the area of the arts. The dancers and creators of this work are Ana Luiza Gomes, Isadora Alonso Faustino, Maria Isabel Torres, Nicolly Lapa and Rafaella Costa.

of live broadcasts and recordings. All these elements dialogued with the movement, in order to modify it as the camera emphasizes highlights points on the body of the performers, which started to become multidimensional and polysemic at the moment when the image of the virtual body was constructed from the materiality of the real and physical body. That is, from the exposure of the real body and material to digital technologies for capturing and recognizing movement, virtual images of these bodies were created, images that, in turn, were inserted and interpreted within the performance, directly influencing the dance work and improvisation of the performers (Fig. 3).

Fig. 3. *Interferências.* Source: Personal Archive. Photograph: Letícia Campos.

One of the ways of creating this virtual body was through the use of a sensor, called *Kinect*[13], which has a built in camera that was activated by the computer and was capable of mapping and recognizing the body of the four interpreters. With it, shapes were projected that responded to the movements, which in turn were also affected by this relationship, since to be captured they had to be within the visual field of the device's camera. Virtual volumes and textures are affected by the dancing body, which in this performance is used as a transmission vehicle (Gatti 2023).

Besides, the public was also invited to be part of the show, with its presence being contemplated in the moments of live filming. This invitation to participate was designed with the aim of establishing a communication that was not unidirectional, in which artists and observers could compose a relationship of exchange and complementarity that would result in different interpretations and reactions to the show, a dynamic contemplated by Gasparini and Katz (2014, p. 55) when they state that: "With regard to communication

---

[13] The Kinect sensor is a legacy product originally released for the Xbox 360 and then the Xbox One series with the Kinect Adapter. Information available at: < https://support.xbox.com/en-US>.

between work and public, it is alive, and continues to develop after a first contact because the flow of information exchange between body and environment is unstoppable".[14]

Public participation took place in moments of overcoming the boundary between stage and audience, in which the exchange between these two spheres took place in passive and active ways, with more or less participation with the spectators. Through danced interactions on the outside of the theater, there were encounters between the dancers and people who were not necessarily participating in the show, which generated a direct and relational contact, in which all reactions became stimuli for the improvisation of movements, checking, then, an unprecedented character to the work, which is transformed according to each established contact.

The cell phone was also incorporated as a scenic element and was used as a materialization of technology within the show, being a necessary tool for the live streamings and recordings to happen. While the scene was taking place on stage and the live filming was being done, the cell phone would turn to the audience, showing the faces of the viewers, which was an indirect interaction between spectator and artist.

Furthermore, phone calls and messages were also sent to people in the audience who had provided their phone numbers when they bought their tickets for the performance. Thus, in another moment, the real and present forms of the body merged with the virtuality, so that both bodies shared space on stage. Santana (2003) dialogues with the idea of an audience with active participation, and argues that:

> In contemporary arts, the plane bends back inwards – and carrying its exterior inwards. Its dimension is three dimensionalized. [...] As a public one is not contemplated, it is observed recognizing itself as an implicator in this observation. (Santana 2003, p. 7)[15]

In addition to being affected by the filming and live broadcasts, the body affected and was affected by the layers of sound. The Movie Guitar application (se já usou antes, trocar por app) was used, as well as other sensors and sound and movement capture softwares, responsible for generating interactive projections with visual effects that responded to movements and music. Access to these elements was present throughout the creation process, which directly interfered in the choreographic dynamics, since it enhanced the dialog between reality and the show's virtual imagery. In this way, dance and music are allowed to be permeated by real events, shaped by the relationship with the technological elements used, such as the video resource, for example, which enables new dynamics to the show, seeing as it allows the time of the scenes to be manipulated in a way that would not be possible in a live presentation. The images used, therefore, could be modified thanks to the resources available, a fact discussed by Santanella (2020), in which the author states:

---

[14] Original text: "No que tange a comunicação entre obra e público, ela é viva, e segue se desenvolvendo depois de um primeiro contato porque o fluxo de trocas de informação entre corpo e ambiente é inestancável." (Gasparini e Katz 2014, p. 55).

[15] Original text: Nas artes contemporâneas, o plano dobrase voltando para dentro – e carregando o seu exterior para o lado interno. Sua dimensão é tridimensionalizada. [...] Como público não se contempla, se observa reconhecendose como implicador nesta observação. (Santana 2003, p. 7).

Formal and formalizable image, but permanently modifiable thanks to the instrument's ability to rapidly encode the representation elements through the successive transformation of parameters. In this multiplicity of possibilities, always reversible, the virtual subverts the register of traditional time, as time that runs and perpetually restarts is constituted by this image. (Santanaella 2020, p. 80)[16]

Creating a dance show that dialogues with technology, both in its conception and in its thematic inspiration, requires a creative search that needs to be shared by the whole team so it can include all the areas. In addition, the dynamic between the different spheres of activity was based in the relationship with the performative environment that shapes how the contact, direct or not, with technologie will be established and with that would be possible to understand how these situations affect and perpass the body. In this way, the body would become action, rather than a performance, because all of these technological stimuli, or not, that interacted with the performers during the show affected and influenced its creation. This work is known as Embodiments and, as explained by Santana (2003), it means that each new experience between the individual and the environment will create new bodies, thanks to that contact with different contaminations.

As a result of the construction of this work emerged a dramaturgy that sought to provoke in the audience a sense of unease linked to questions about today's lifestyle, which revolves a lot around contact with technology.

Interferências provokes a sense of restlessness by representing this feeling through the dance of the performers and the use of sound, digital and visual elements during the performance. By combining music, dance and technology and exploring the possibilities of creation that derive from this combination, the show performs a multi- and interdisciplinary work, which allows the audience to be exposed to a variety of meanings, including the daily restlessness of a reality highly influenced by technology.

The creative work involved personal - and at the same time collective - exercises to interpret the meanings that each of the performers attributed to their relationship with technology. The search for experiences that would inspire productions for the show resulted in various ways of embodying the positive and negative sensations related to technology, such as anxiety, agitation, comfort, boredom, immediacy and the restructuring of relationships.

Added to this, musical, visual and digital interpretations of these sensations are also explored. As previously mentioned, from the use of apps and software for recognition and replication of movement and music production, in addition to live music, a process of "translation" of sensations to musical rhythms and projections of images, videos, textures and colors is carried out.

In a summarized manner, and assuming that immersion in the digital reality is a process shared by many people, it was possible, through the development of all the processes stated above, to provoke a feeling of recognition by the public of the elements

---

[16] Original text: Imagem formal e formalizável, mas permanentemente modificável graças à capacidade do instrumento de codificar rapidamente os elementos de representação através da transformação sucessiva de parâmetros. Nessa multiplicidade de possíveis, sempre reversível, o virtual subverte o registro do tempo tradicional, pois o tempo que corre e perpetuamente recomeça é constituído dessa imagem. (Santanaella 2020, p. 80).

and sensations present in the daily life of modern lifestyle. This lifestyle generates consequences that are felt in our bodies, which face a relationship of mutual modification with the environment (Santana 2003).

It can therefore be said that technology is experienced and felt by our bodies everyday and, just as it has an impact on artistic creation, it also has an impact on other aspects of daily life. Therefore, as dance artists and teachers, after going through the creative process of the show, we were instigated to question more about the impacts of technology in an educational context and ways of applying these new possibilities for dance research in teaching spaces beyond the university, in order to reach an audience that doesn't necessarily study dance and understanding that technology in the classroom restructures the traditional teaching habits and generate both positive and negative changes.

3 Dance and New Technologies Beyond the University: Teaching Techniques in Different Spaces

The integration of technological resources in dance has proven to be a great strategy for expanding creative possibilities and pushing the limits of teaching and learning. When applied in a teaching context, in schools and classrooms, technology also acts as a great ally to facilitate students' connection and understanding of dance, especially when it comes to the context of basic education schools, where the youngest generations of students have grown up in a digital environment and are already familiar with technological interactions. It can't be ignored, however, the penetration of the internet in today's life and, by analyzing this insertion only through the lens of dance, it's possible to say that this influence brings to people a new access to dance and networked experiments (Bittencourt 2021). Therefore, using technology in teaching means incorporating a language familiar to students and making the learning process more accessible and engaging. This issue has already been highlighted by Bittencourt (2005), when she talks about the importance of studying dance from this new perspective:

> The issue of dance education through new media should also be considered in Brazil so that future professionals are not afraid of real, or virtual, organic or silicon bodies, but that they can study dance from another perspective. It is seen that the readaptation of man with technology and art is necessary, including in education, since time and space via the virtual network are different from the time and space of body movement, which we call real. (Bittencourt 2005, p. 11)[17]

Analyzing the technological equipment available today, it was questioned possible ways to transmit and expand the understanding of technology for students and what resources could be taken to a school. From this, some challenges were perceived. Among

---

[17] Original text: A questão da educação da dança através das novas mídias deve ser pensada também no Brasil para que os futuros profissionais não tenham medo dos corpos reais, ou virtuais, orgânicos ou de silício, mas que possam com isso, estudar a dança sob outra perspectiva. Vê-se que a readaptação do homem com a tecnologia e a arte é necessária, inclusive na educação, já que o tempo e o espaço via rede virtual são diferentes do tempo e do espaço do movimento corporal e que chamamos real. (Bittencourt 2005, p. 11).

them, technical limitations and technological failures, such as the lack of equipment availability and internet connection, in addition to the difficulties in fluidly integrating technology into dance and music work, so that it becomes an addition to the teaching and not a distraction.

This research was carried out with students from two basic education schools[18], one private and the other public, and from a dance school[19], all in the city of Campinas-SP, with beginner dance students. All the strategies used in the classroom emerged from the previously mentioned performance experiences: the performance at the Art Gallery and the *Interferências* spectacle. The objective was precisely to investigate whether the same technological devices used in a performative environment could be worked on in a dance teaching context.

The age of the students from the private school was between 6 to 13 years, the students from the public school had between 10 to 11 years and the students from the dance school had 7 to 12 years. During this experience, that took around 4 classes[20] in each school, 60 children participated, and the work was conducted in such a way as to provide greater autonomy to the students, so that they could explore their limits, interests and imprint their identity in the movements in a way they could identificate with their dance. We've noticed that these classes could be applied to a very large rage of ages, but some adaptations were necessary. One example of a very important adaptation is how the explanation of the exercises were transmitted to the students. Younger kids needed more ludic instructions, for example, instead of saying "feel the difference between heavy and lighter movements", the language used was "move your body like it's made of stones" and then, "move your body like it's made of air". And, in addition to create those instructions, the technology was applied. The students could be instigated to change their movements and to create sounds as their body moved, and later they could be instigated to recreate a movement from some sound they liked the most.

This "formula" of combining traditional elements of dance (like, weight, speed, use of space), with the use of technological resources, was applied to all ages. This was made, in order to improve the learning experience of students, who despite having contact with technology, had not used it creatively in dance classes. As a result, an increase in students' interest in the class was observed, but certain eventualities and limitations arose that required a contingency plan in place. Not all activities could be applied in all schools, we had some infrastructure issues that limited our options.

All schools had sound boxes available and, based on this resource, dance teaching with musicalization was carried out through dynamic activities that worked with rhythm and perception. The use of technologies in teaching and learning was guided by the provision of tools by schools that could be used in the classroom. It is important to emphasize the need to think about the musical work during dance classes:

---

[18] The schools were: "Raul Pilla" and "14 Bis".

[19] "Instituto Arnea" is the name of this dance school. It is a specialization school with no fees to students, therefore all the students can study dance without paying any fee.

[20] It was a month of work especially focused on the development of dance in conjunction with technology. It is important to highlight that dance work in all these classes had already started since the beginning of 2023, however with other study focuses.

One cannot talk about dance without thinking about the relationship between space and time, which are inherent elements of this language. In this sense, we noticed that the musical work in dance classes allows the student to develop more temporally, acquiring gains in relation to the rhythmic sense. (Faustino 2022, p. 10)[21]

Mixed reality was one of the proposals worked on. In it, the students jointly developed a choreographic cell (a composition of sequenced dance movements) and remodeled it with rhythmic variations and different spatial placements. The class was divided into two groups that had to dialogue with each other bodily. One of the groups was being filmed and was outside the room, so the only access to communication would be through a live video transmission, which was being projected to the group inside the classroom. Through this creative and playful exercise, students were able to work on a new spatial and social perception. This dynamic however was interfered by the internet connection, which caused delays in the images and impacted the dialogue. Despite this, it was possible to unite the group and use these "failures" in favor of the dynamic.

Nevertheless, only in one of the places it was possible to use the live transmission application, *DroidCam*, since it requires a place to transmit the image for use, such as a projector or a computer, and only in one of the schools these materials were available. In the other two schools, the cell phone was the device that centralized the classes. It was the most accessible technology to work with, in addition to being very present in the students' lives. With this, it was possible to reframe their functions, which are usually linked to everyday needs. For example, it was proposed the use of the camera in dynamics for creating video dances, which were later used to publicize schools or as interdisciplinary material that would be worked with teachers of other subjects.

Furthermore, students were introduced to Move Guitar and, with an almost innate drive, quickly learned to use the app. One of the main obstacles to the task is the fact that, as it's still under development, this is not an application available for all devices, which limited its use to only one that was provided to them. Added to this, because the cell phone was held by the hands, the students initially restricted its movement to the arms, without investigating ways to produce the sound through more complex movements. At the same time, as a result of the individualization caused by cell phones, students were focused on their own research, with little affective and social development.

Therefore, it was necessary to bring proposals that integrated all the children, as is the case of cooperative exercises and work with eyes closed, in which the guidance was mostly through sounds, verbal cues and mental images. In other words, they depended on short verbal information that directed the focus of attention to important aspects of the movement or on imagery representations (real or metaphorical) that helped in understanding the activities. These conductions were sometimes made by the students themselves, in order to stimulate sociability and autonomy in dance.

---

[21] Original text: Não se pode falar em dança sem pensar na relação espaço e tempo, que são elementos inerentes a esta linguagem. Neste sentido, percebemos que o trabalho musical nas aulas de dança possibilita que o aluno se desenvolva mais temporalmente, adquirindo ganhos em relação ao senso rítmico. (Faustino 2022, p. 10).

In the last week of research, it was also possible to carry out another proposal with technology in the private basic education school. The school was interested in the proposed interdisciplinary creation between dance, sound and technology and accepted to provide another space for one of the classes, which had a white wall, ideal for working with shadows created by scenic lights (see Fig. 4). On this last day of activities, the students had already created a choreographic score based on the sounds and were able to delve deeper into this investigation using the lighting. This activity later developed into a very similar exercise that was applied in the work "Interferences" (see Fig. 5), which consisted of using two reflectors so that the shadow created was double, which generated a three dimensionality to the image. This exercise aroused a lot of interest in the students, who constantly wanted to test new ways of moving and creating shadows, creating postures in pairs and groups, which were later explored in another month of activities.

Fig. 4. Students working with shadows and dance. Source: Personal archive.

A difference observed during the application of the proposal was the work with technology in basic schools and in dance specialization school, since the target audience of students varies between these two scenarios and, therefore, deals with classes in different ways. It is important to find a balance between technology and body movements that creates a coherent and engaging narrative with the context in which it is situated. In basic education, for example, students will not necessarily be interested in dance, since their curriculum will be broader. Therefore, the activities must be very dynamic and must stimulate different areas of movement, in order to lead the student to try new movements, thus stimulating creativity and motor coordination without insisting too long on the same proposals. In specific dance training schools, activities can be more in depth and precise to provide greater technical evolution. In both cases, teaching strategies combined with technology are excellent allies to expand students' creative possibilities and help them develop their artistic expression in innovative ways.

Fig. 5. Shadows in *Interferências*. Source: Personal Archive. Photograph: Leticia Campos.

4 Conclusions

The collaboration between dance, music and new technologies, which was analyzed in the previous paragraphs, showed that the interdisciplinarity between these areas of knowledge makes it possible to expand the boundaries of artistic expression and creates innovative experiences with great creative potential, whether in performative contexts or in teaching and learning contexts. The technological resources and methods applied in the classroom or in shows and performances will vary according to the contexts, but it is important to emphasize that mobile devices, such as cell phones, democratize access to technology and allow for new opportunities for the development of approaches that reformulate dance work.

By analyzing and observing the history of the use of technology in dance, it is clear that the evolution of the presence of digital in everyday life has transformed human behavior. It is essential, therefore, that teachers and artists are willing to work on an inter-communication between technology and dance to discover ways to use them together, and in the best possible way, to achieve the objectives of each project.

Within the two main topics exposed in this article: dance shows and dance teaching in classrooms, in which the representation and use of technology were analyzed, traces of similarity can be observed. Even with totally different proposals, when exposed to the use and practice with technology, these two contexts obtained similar results.

The use of the *Movie Guitar* application, for example, was a technique covered by both topics. The shared use of this technique shows the flexibility and versatility provided by this method which, both in *Interferências* and in the schools, was used as a tool to boost improvisation in dance.

For instance, the dance performed in the show was a language familiar to the performers and, therefore, the sounds created only responded to the movements made from improvisation. In addition, in that context, the sounds and dance dialogued with the projections in a scenic and performative way, which was fundamental for the narrative construction of the show. In schools, as a pedagogical practice, cell phones and apps were also used in dance improvisation contexts. However, since the children did not have a very varied repertoire of movements, they constructed their movements based on the app's sound stimuli, and not the other way around, as happened *in Interferências*.

In addition to the issue of improvisation that was shared in both topics, the high connection between everyday life and technology was evident. This link was explored in the spectacle and the schools through artistic representations of the effects of technology immersion in modern society and by the intimacy presented by students, when exposed to cell phones and applications respectively.

However, with regard to the educational context, even if it is an interesting and useful tool for improving dance teaching, technology should not assume an unique and exclusive role in the teaching process. It's necessary to avoid excessive dependence on it, in order to not replace face-to-face interaction between students and teachers with virtual interaction. Dance is a form of corporal and artistic expression that requires human connection, emotional and physical expression and the excessive use of technology can dilute these elements, which are fundamental and necessary for the development of this language.

Finally, this work and the raised applications act as an active process of thinking and acting in the face of a reality highly influenced by technological advances and their growing presence in the social spheres that permeate and transpose the routine of individuals. In an effort to study, understand and produce from and through this reality, artistic and pedagogical processes enter as tools and responses to deal with a technological world that creates new forms of socialization and relationship, in a continuous and dynamic way, besides to changing the way you deal with everyday life.

References

Bittencourt, A.T.: A influência da tecnologia na dança. Coletânea de Arquivos do LABLUX Laboratório de Iluminação da Unicamp (2005)

Gasparini I., Katz, H.: COMUNICAÇÃO ENTRE DANÇA E PÚBLICO: o papel do coreógrafo na construção da relação obra-espectador. DANÇA: Revista Do Programa De Pós-Graduação Em Dança 2(3), 51–66 (2014)

Gatti, D.: De uma margem a outra: trabalho composicional multimodal entre movimento, som e tecnologia. In: X Congresso da ABRACE, 2018, Natal RN. Anais X Congresso da ABRACE 2018, vol. 19. pp. 1–12 (2018)

Gatti, D.: Dança em redes de saberes no ensino superior: o papel do Trabalho de Conclusão de Curso Artístico. Editora CRV (2023)

Fasutino, I.A.: Dança com musicalização para crianças. In: 22° Conic Semesp, 2022. Anais 22° Conic Semesp (2022)

Menicacci, A.: O ensino da dança face à tecnologia digital (2004). http://idanca.net/o-ensino-da-danca-face-a-tecnologia-digital/. Accessed June 2023

Postman, N.: Tecnopólio: a rendição da cultura à tecnologia [Technopoly: the surrender of culture to technology]. Nobel, São Paulo (1994)

Santaella, L.: NÖTH, Winfried. Imagem: cognição, semiótica, mídia. Iluminuras, São Paulo (2020)

Santana, I.: Imagens do corpo através das metáforas (ocultas) na dança-tecnologia. INTER-COM – Sociedade Brasileira de Estudos Interdisciplinares da Comunicação. XXVI Congresso Brasileiro de Ciências da Comunicação – Belo Horizonte (2003). https://www.academia.edu/download/30638253/2003_np07_santana.pdf. Accessed 12 junho 2023

Santana, I.: Corpo aberto: Cunningham, dança e novas tecnologias. Educ, São Paulo (2002)

Vaughan, D.: Merce Cunningham: Fifty Years. New York (1997)

Dance and Technology: Different Readings of the Technologically Mediated Body

Diogo Angeli[✉] ⓘ

Universidade Estadual de Campinas, UNICAMP, Campinas, Brazil
dangeli@unicamp.br

Abstract. This article investigates the relationship established between dance and technology in the contemporary artistic-creative scope, with special focus on the different poetic developments present in the body and dance in virtualized contexts. Such reflections seek to understand and deepen their modes and structures of creation, the different relationships established between movement, body, and technology, as well as their receptivity and interactivity with the public and / or spectator. As a methodology, the article focuses on the production of knowledge from the investigation and reflection on theoretical research developed in the area in conjunction with the practical analysis of works relevant to the proposed discussions. To this end, the research is based on theoretical studies on dance-technology (SANTANA, 2006; AMORIM, 2009) and screendance (ANGELI, 2020 and 2022; CALDAS, 2012), in addition to proposing the analysis of two Brazilian artistic works produced in this creative category and important for the reflections on dance and technology: the screendance *Unusual* (2018) and the artistic work *Interferences* (2022).

Keywords: Dance · Technology · Creation

1 Introduction

In the artistic field, the production of artistic works mediated by technology has become increasingly frequent, showing itself as a territory with great potential for expansion today, in which dance is included.

The dances are contaminated by stimuli present in their environment and, together with technology, find different ways of acting, causing ruptures in the boundaries between face-to-face and virtuality. As pointed out by Carvalho; Pronsato (2020), digital culture and technological mediations present in contemporary artistic productions have transformed the ways of thinking and doing dance, providing new ways to look at the body and movement and provoking transformations in spatial and temporal relations.

Initially, this article brings the reader a brief theoretical framework on the relationship between dance and technology from an immersive point of view, that is, considering that such languages are contaminated when placed in a state of creation, to the point of reconfiguring their structures. In this way, the body mediated by technology is also

A. L. Brooks (Ed.): ArtsIT 2023, LNICST 564, pp. 107–119, 2024.
https://doi.org/10.1007/978-3-031-55319-6_8

highlighted in the discussion, precisely because it is subjected to a series of stimuli and interferences cap0able of transforming its qualities and abilities, enabling the creation of new conditions or expressive states, as well as offering new readings from its virtualized condition.

Finally, the article presents the analysis of two Brazilian artistic works that explore the relationship between dance and technology, the screendance production *Unusual*[1] (2018), created and directed by Diogo Angeli, and the artistic work *Interferences*[2] (2022), created by Ana Luiza Gomes Przsiczny, Isadora Alonso Faustino, Maria Isabel Torres Dos Santos, Nicolly Karoline Moreno Lapa, and Rafaella Ferreira Costa. The chosen works dialogue with the purposes of this article by presenting situations and creative explorations of the body in dance and technology interesting to the discussions promoted.

Thus, the article seeks to investigate the relationship established between dance and technology in the scope of creation in dance in contemporary times, establishing a dialogue between the theoretical material and the practical productions in the segment, seeking to understand the possible consequences provoked to the body and dance in this type of creative condition.

2   Dance-Technology, an Immersive Context

Dance-technology is understood as the creations in dance mediated by technological devices, which promote the approximation between dance and digital culture (SANTANA, 2006). This approach can be explored in dance creations from different formats and intensities, in face-to-face, virtual or hybrid contexts, and may involve interactive, telematic, inter media, software, among other technological elements.

From the technological mediation, dance, body, and movement are seen under new frameworks and dimensions in relation to their poetic-expressive conjuncture, in the same way, different meanings are created from such readings and interpretations, because, once transformed into communicable media, different modes of perception and communication are created, which modifies the connections established between such elements with the environment and the spectator.

In this spectrum, it is understood that the relationship established between dance and technology is not limited to a simple overlapping of means or languages, but is an immersive relationship, permeated by transdisciplinary exchanges in which one medium is affected and transformed by the other, both regarding its structure and composition, as well as its communication, as Amorim corroborates:

> Dance in interaction with digital media does not mean a composition format exhausted of human emotions and subjectivities. The electronic medium constitutes a new way of exposing the work and establishing a new dialectical relationship between the author and the spectator. This new environment must remain aware of its power to stimulate creativity and the entrepreneurial spirit and enhance its social applications as some of the building functions of the values essential to human development (AMORIM, 2009, p.1, our translation).

---

[1] Original tittle: *Insólitos*.

[2] Original tittle: *Interferências*.

Santana (2006) considers that the bodies and movements present in dance from the technological mediation are constructs of an expanded thought that do not dichotomize mind and body, nature and artificiality, reality, and virtuality, since they are seen as bodies in transformation, modified bodies, in the form of communicational media that are placed in constant exchange with the cultural and social perceptual environment. It is in this perspective that the environment-individual relationship is understood from a mutual implication, that is, technology – also a creation by humans – is seen as the result of an evolutionary path, that is, technology becomes inserted in the existential actions of the subject, transforming elements, means and relations.

Thus, the body is contaminated by the elements present in the environment in which it inhabits and dance with technological mediation emerges from these relations, resulting in the approximation of the body/dance with the digital culture. This approach reconfigures the body and technology, being understood no longer as complementary elements or in a state of relation, but as a new structure. This restructuring of the body creates hybrid products, reconfigured and armed with their own laws and principles, emerging from this virtualized state between dance and technology, which are configured in different formats. Such creations allow multiple readings of the world through dance, each with its specificities as a language, and this structure – dance/technology – has the potential to create different stimuli, bodily or technological, transforming the perceptions of the body and dance, broadening the notions of time and space, creating new relationships between organic and non-organic elements and with the perception and interactivity of the spectator.

There are numerous paths and possibilities of approach and interaction of technology with creative experiments in dance, which can be carried out in face-to-face or virtual contexts. Among them, we can mention:

- The use of technology associated with the creation of new expressive languages that have hybridity as a primordial condition of their nature, as is the case of screendance, a hybrid language that integrates knowledge from dance, cinema, and audiovisual.
- The use of technology as a tool of support and provocation in creative processes in dance, making it possible to experience other ways of understanding the body in movement through virtuality, generating new possibilities and creative experiences. The use of software for the creation and virtual manipulation of bodies and movement[3] allows the exploration of such potentialities.
- The use of technology as an integral poetic component of artistic phenomena in face-to-face contexts. The use of cameras, images, sensors, projections, among other technological devices can integrate the face-to-face presentations to change the perceptions of the spectator in relation to the body, movement, space and poetics of the

---

[3] In this way, one can cite software such as *Lifeforms* (today also known as *Danceforms* 1.0), a software that uses avatars for the creation of dance movements and *Poser*, another software that makes use of avatars in the creation of dance movements with possibilities that go beyond human anatomical abilities.

work. Telematic and interactive performances, for example, use technology to manipulate the body, movement, and space in real time, making use of sensors, software and programs for projection and composition of images[4].

Technology, in this context, is understood as an idea or thought that is incorporated into an experience in dance with technological means, distancing itself from the simplistic idea that considers such an experience only from the use of technological artifacts, employed in a creation with such purposes. The intention is to make such devices act as mechanisms that manipulate information and create new methods, forms, and thoughts of creation in dance, a mediating interface between man's thought and technological means, since "virtuality only means another possibility of existence" (SANTANA, 2006, p. 110, our translation).

Thus, technology stands as a means that enables the construction of new relationships between artistic work, author, and spectator. Venturing into this universe requires the dance professional to develop new skills and technical and creative capabilities about the technological environment and its relationship with the processes of creation in dance, building new visions for the body and movement from its virtualized configuration.

3   The Technologically Mediated Body

When technology is used as a mediator in the creative processes in dance, the body also has its habitual creative condition displaced, becoming a means of communication, a media in itself. In the creative context, the body is crossed by various information, technological or not, and the information and technologies that make up such an experience, also become an integral part of the body. In the case of creations with the presence of technology, the information exchanged between the body and the technological environment alters their qualities and perceptions, providing that new skills and different forms of interpretation are achieved in this condition.

In virtuality, for example, the body finds different possibilities for the realization of its actions that allow it to expand its physical limits, that is, to break with the natural limits established by time and space. In relation to time, new speeds and cadences can be provided to the movement in virtuality, as well as new skills and capabilities that go beyond the boundaries of the "possible" in the face-to-face plane, such as gravitational changes, temporal displacements, accelerations and reductions in the speed of their actions, among others. With regard to space, in virtuality it becomes possible to create temporal windows that bring together distinct spaces or provoke the multipresence of the body. Thus, technology as a mediator of the body in the creative act, allows to expand the perceptions of the body, expand its relationship with time and space, as well as change

---

[4] Examples of real-time image composition software include: *Isadora*, a software created by Mark Coniglio and Dawn Stoppiello that enables interactive performances in real time. From a camera attached to the computer, the software allows you to identify the movement and modify it, applying different effects to the captured videos; *Active Space* is a media collection system, organized in a network and in real time, that allows the creation of interactive environments from motion tracking technologies, carried out from cameras that capture movement. The material is processed by the system that enables the organization and projection of these images in several layers, being a resource widely used in shows that make use of telepresence; among others.

its expressiveness and understanding, providing new perspectives and relationships with physical, aesthetic, social and cultural contexts from its manipulated condition.

As a subject, the virtualized condition of the body can establish different perspectives and relations with the spectator's reality, with multiple readings arising from the same context. The body, when mediated by technology, presents itself as a type of media, with different possibilities of organizing information because, in the condition of image, the body finds other ways of acting, thinking and expressing messages.

Once body and movement are transformed into images, visual perceptions and imagery representations begin to compose their meaning. In this situation, Peirce (2005) emphasizes that the reading of an object from its image, triggers a set of associations and interpretations – signs – that remain connected to the form of the elements represented and to the sensitive universe of its observer.

> A Sign, or representation, which refers to its object not so much by virtue of any similarity or analogy with it, nor by the fact that it is associated with general characters that this object happens to have, but rather by being in a dynamic connection (spatial inclusive) both with the individual object, on the one hand, and, on the other hand, with the senses or memory of the person to whom it serves as a sign. (PEIRCE, 2005, p. 74, our translation).

The insertion of the body-image in the artistic experience expands the poetic possibilities of the body, allowing the construction of different expressive relations from the imagetic and technological universe that, in dialogue with the studies of semiotics, also expand its perception and reading. Thus, new metaphors can be created due to the experiences and transformations resulting from the interactions of the body with the technological universe.

The vagueness inherent in the concept of sign is responsible for establishing multiple connections, amplifying their meanings due to their subjective nature. The composition of the images of the body considers the stimuli and references present in the work and external to it, creating signs that allow to potentiate their perceptions, promoting different readings and metaphors due to the imprecision present in its structure. This condition allows the body mediated by technology to find different ways of expressing or defending poetic, artistic, social, cultural, or political points of view. Technology, including computers, videos, projections, sensors and images, allows to reproduce, and amplify the symbolic productions created by the body and dance. By allowing hybridization between these languages, it is possible to modify and incorporate new references to dance, as well as explore real and virtual elements in synergy. The construction of signs in a dance with technological mediation, in addition to carrying such universes in its composition, also involves the viewer's perspective in this equation, because the interpretation of their meanings passes through the universe of the recipient of the work, bringing together the medium and subject, which allows to evidence social, economic and political aspects present in this reality. In this way, the understanding of technology in conjunction with the concept of sign proposed by Peirce offers new perspectives to understand the body as a media that interacts with life, the social world and culture.

In this way, technology is seen as part of the process of creating the body, and not as the result, because its presence makes it possible to organize creative information in different ways, integrating the intentions of creators, by exploring different ways of presenting their poetics, with the desires to construct meanings and deal with creative complexities. The transformations that occurred in this case are related not only to the body, but to the structure of the artistic work as a whole, since the technology is developed together with the environment, modifying it and promoting different connections.

3.1   Technology in Favor of Narrative in Screendance *Unusual*

Unusual[5] is a screendance produced in 2018 in Brazil and directed by Diogo Angeli. Its creation was part of the author's master's thesis developed at the State University of Campinas, UNICAMP-Brazil. *Unusual* has as it theme gender identity, specifically bodies that are non-binary. It brings together through images, actions, narrative contexts and time-spatial manipulations of the body and movement, elements that enable different readings of the body to the spectator, questioning the decriminalization, conflicts and violence practiced to these bodies in society, in addition to exploring other poetic models of dance.

The language of screendance represents one of the possibilities of exploration between dance and technology, because as pointed out by Caldas, "new technologies do not stop tensioning dance in the direction of a reinvention" (CALDAS, 2012, p.240, our translation). The poetics of screendance brings together elements from dance, cinema, audiovisual and technology from a unified, interrelational and hybrid dialogue. Its creative process provides differentiated poetic conditions to dance, distinguishing itself from face-to-face productions. Once virtuality is established in the process, new ways of thinking about the body and movement can be produced, transforming the references of time and space, and finding different possibilities of expression and reception. According to Caldas (2012), it is from the manipulations of time and space in virtuality, from the ability to discontinue and resize them, that dancing the impossible becomes possible.

The virtuality present in this context is associated with the technological universe, linked to cyberculture and digital technologies, and, at the same time, connects to its creative and transformative potentiality, as emphasized by Levy (2001) when presenting the virtual as a transforming force, a creative power, capable of generating new modes of production and realization, that is, a medium in a constant state of creation.

Virtuality, therefore, is present in screendance from the elements that compose it as a language – such as the camera, the screen, the images, the editing process and the technologies involved –, as well as in its creative process and in the final format of its productions – the video –, influencing the relationships that the artistic work establishes with the audience, with space, with time and the other elements that relate to it. Thus, the creative process of screendance involves face-to-face and virtual moments in relation to the explorations carried out with the body and movement.

In this way, it can be verified in *Unusual* (2018) that the exploration of the body through virtuality promotes different expressivities to the work, constructed, in parts, from the imagetic signs and the time-spatial variations of the body and the movement in this context. About the dance-technology relationship, it can be highlighted that the virtualized exploration of bodies in screendance and the relationships established between

---

[5] Screendance available at: https://youtu.be/U7-B0rCN7QM. Access in: 09 jul. 2023.

dance, images and imagery signs present in the work, alter the process of construction of movement in dance and the perception of its spectators from audiovisual technology. The body seen on the screen in *Unusual*, sometimes presents itself from temporal and spatial distortions, under qualities that extrapolate the physical limits of the body as matter, this due to the intentions and creative and expressive challenges present in the artistic work that, as stated by Angeli, in addition to reflecting on the theme of gender, "one of the challenges I embraced for the screendance *Unusual*, was to reconstruct the movement, expressiveness and narrative of the work, through the editing process and the time-spatial manipulations of the video" (ANGELI, 2020, p.133, our translation) (Fig. 1).

The manipulations applied to the bodies in *Unusual* were only possible due to the relationship established between dance and technology that, when explored in the work of screendance, brought new references for the process of creation in dance and for the reading of the body and movement by the spectator.

Technology as a mediator of the bodies that perform in *Unusual* interferes in the creative process in dance. In screendance, unlike dance productions performed face-to-face, there is an interaction of the body with different technological devices during its creative stages, such as the camera, computers, and image editing software. Such changes promote new paths for creators considering the dialogue between body, image, and technology. In *Unusual* these creative possibilities made it possible to establish different ways of relating time and space with the body and movement in virtuality, enabling the duplication or multiplication of bodies in space, making them occupy different places at the same time, as well as allowing several bodies to occupy the same space. It also allowed the realization of manipulations and physical changes in the bodies, transforming their compositional organization and creating variations of their physical structure as if, in some moments of the video, they did not have bones or carried limbs and body parts of disproportionate sizes. In addition, different velocities could be explored from the accelerations and decelerations of their actions.

The mediation of the technology present in the bodies in *Unusual* also modified the perception of the spectator in relation to the artistic work. The technology together with the imaginary universe explored by screendance, promotes new dialogues between body, image, and narrative. The presentation of the work in video format dialogues with the knowledge of semiotics that considers that "image is a message made up of iconic signs" (COVALESKI, 2012, p.89, our translation) and that sign "is that which, under a certain aspect or way, represents something to someone" (PEIRCE, 2005, p.46, our translation), that is, the composition of the images in the screendance – the organization and the relations established between the narrative elements in the imagetic plane – promotes the creation of different meanings to its receiver from the reading of these images, which present meanings connected to their visible content – their form – and their subjective and imaginary content. Regarding the reading of the images, Santos states:

> The association of planes in a given sequence creates in the mind a meaning, which is determined by the verification of their connection, creating a need to mediate each plane and its relationship with the linear arrangement of facts/frames. Thus, this imputed order (one reads intentionality), from one plane to another,

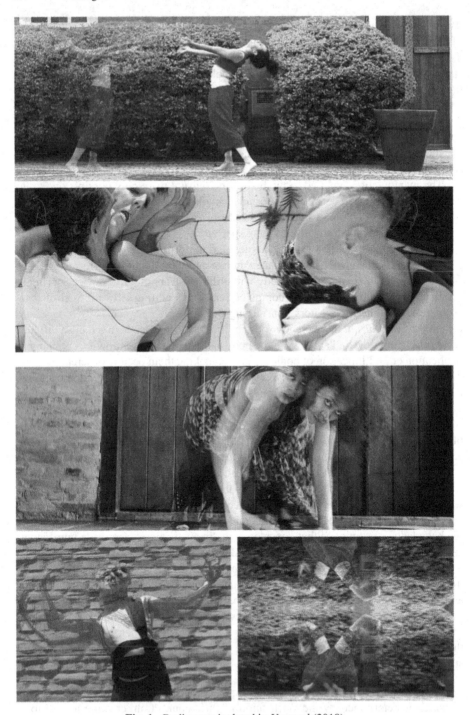

Fig. 1. Bodies manipulated in *Unusual* (2018).

creates an interpreter responsible for causing the mind to infer and recognize these connections in a diagrammatic way and produce (effect) an understanding of the whole, or area of information. (SANTOS, 2013, p. 3, our translation).

With each different choice made in the combination between the images that make up the work in screendance, new ideas, looks or perceptions are also suggested to the viewer. Thus, Angeli adds that:

The association of the narrative elements, created from the editing and the imagery condition of the screendance, provokes[6] in the mind of the spectator different ideas that expand their interpretation, since, when transformed into images and signs, the narrative elements create new perceptions and combinations, generating different layers of reception that articulate with the mind and the capacity of understanding of those who observe them in their totality (ANGELI, 2022, p.149, our translation).

In this way, the relationships established between the bodies and the theme proposed by the artistic work with the spectator were also affected by technology. The grotesque, the deformations and the multiplication present in the bodies in the screendance *Unusual* suggested different relations and perceptions to the spectator from their imagetic and manipulated condition, being at the discretion of the spectator to establish the relations between the images and their social universe, as well as to build the relationship of these images with gender identities and expressions and non-binary bodies. The deformation explored through the technology on the bodies in *Unusual*, for example, suggested their relationship with monstrosities, and in the thematic context about gender, such monstrosities could be seen as a prejudiced and discriminatory view placed onto to those bodies.

3.2 *Interferences,* **Experiments Between Sound, Body, and Technology**

Another experiment with bodies mediated by technology is present in the dance work *Interferences* (2022) created by students[6] of the undergraduate course in dance of the State University of Campinas – UNICAMP / Brazil in partnerships with other artists on music and multimedia, advised by Professor Daniela Gatti.

Interferences proposes to integrate the languages of contemporary dance, music, new media, and technology, exploring possibilities of relations between body and technology, from the intervention of motion capture cameras in real time, telematics, image manipulation and projection software, motion reading sensors, among other experiments related to sound creation and scenic lighting. From the analysis of the body in this technologically mediated context, it becomes possible to visualize different creative strategies adopted by the creators in conjunction with technology, as well as different paths and poetic readings about the work.

---

[6] *Interferences* (2022). Created by Ana Luiza Gomes Przsiczny, Isadora Alonso Faustino, Maria Isabel Torres Dos Santos, Nicolly Karoline Moreno Lapa and Rafaella Ferreira Costa. Musician-composer: Eduardo Koji Takeshita Junior. Imagery creator and responsible for interactive projections: Guilherme Zanchetta.

116   D. Angeli

Among the experiments in dance and technology in the artistic work, some moments can be highlighted to broaden the discussions proposed by this article.

With regard to the exploration of space, *Interferences* combined real and virtual references, made use of telematics in the composition of the scene, experimenting with remote creations with multiple angles of the same space inhabited and revealed by the performers bodies, bringing to the audience different perceptions in relation to the internal and external spaces of the place of presentation, as well as about its face-to-face and virtual condition, questioning and reflecting on the potentiality of dance from the exploration of technologically mediated space.

Through cameras, positioned sometimes outside the performance venue and others in the scenic space itself, *Interferences* provided the audience with the interaction of face-to-face bodies and virtual bodies. The synergy present between the two spaces, face-to-face and virtual, established different relationships between the spaces, performers and audience, expanding the creative limits of dance from telepresence and instigating the audience to perceive and reflect on the creative potential of dance in virtuality and on the characteristics and differences present in such spaces, both in relation to their physical aspects, artistic, social, and cultural (Fig. 2).

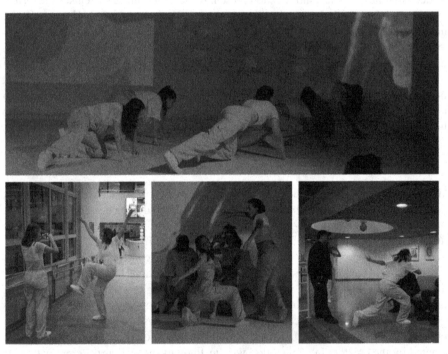

Fig. 2. Exploration of Face-to-face and Virtual Space in *Interferences*

The artistic work also used motion sensors and the projection of images. More specifically the *Kinect* motion sensor and the *TouchDesigner* software that enabled the capture, manipulation, and projection of images. For the interaction of bodies and the creation of imagens, the show experienced different possibilities of relationship between the interactivity of bodies and the composition of images.

The imagery projection present at the bottom of the stage, conceived from the junction of a multitude of small circles that together formed an abstract figure, was manipulated through the *Kinect* sensor and the *TouchDesigner* software. The sensor captured the movement of the performers on stage and, from the body stimuli recorded and the programming pre-established by the *TouchDesigner*, changed the background image, causing the circles to move according to the movement of the performers on stage. This experiment presented different possibilities of interaction between bodies and technology, as if bodies and images could dialogue choreographically, establishing a condition of creative partnership (Fig. 3).

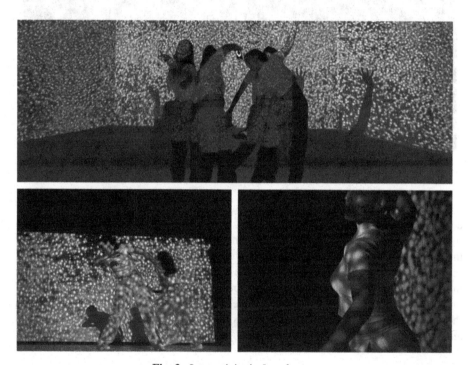

Fig. 3. Interactivity in *Interference*

The relationship among sound, bodies and technology was also explored during the presentation. The artistic work had the participation of a musician-composer, responsible for creating the live soundtrack, which was technologically modified by computer programs and by sensors and motion applications installed on the performers' cell phones (Fig. 4).

Fig. 4. Sound, Bodies and Technology in *Interference*

The show explored the relationship between sound, bodies, and technology under different situations. The *Movie Guitar*[7] app, installed on the performer's phones, interfered in the music track, through the movement of the cell phones. *Movie Guitar* used the motion sensors in the cell phones to interfere in the soundtrack. In the scene, the movement of the cell phones was read by the app and transformed into sounds, allowing the creation of sound interference from the movement of the bodies that carried the cell phones. In addition, the music was also connected with the production of images through the *TouchDesigner* software, creating multicolored imagery landscapes from the sounds and movements produced by the musician.

4 Final Considerations

The creative paths trodden by dance in contemporary times have increasingly assumed interdisciplinarity as part of their explorations, that is, the availability to establish different types of relationships and experiences with other areas of knowledge, in which technology is present, influencing their creative processes and methods. In the examples analyzed by this article, it was possible to verify such an approximation and deepen such questions.

---

[7] The application was created by Jonatas Manzolli, a professor in the music department of UNICAMP.

The movement and the body explored from the technological mediation in *Unusual* and *Interferences*, made it possible to integrate different paths to the creations in dance. In the *Unusual* screendance, the presence of the camera and the editing of the images incorporated new creative possibilities for the dance, adding virtuality as part of the creation process, with the ability to transform the material worked face-to-face. These strategies allowed for the distribution and exhibition of this screendance in the 15th Sans Souci International Film and Dance Festival in the USA, with screenings in the cities of Boulder and Lafayette in Colorado, USA (2018); 12th São Carlos Sreendance Festival (2018) in São Carlos/Brazil; 12th Mostra Curta Audiovisual in Campinas/Brazil. In addition, it allowed to direct the dance to new circuits of presentation, from the screen and cyberspace and, with this, also to add new audiences. Such explorations can also be seen in *Interferences*, whose process explored multiple relationships and readings of the body from the remote context, in dialogue with devices, sensors, and software that transformed its relationship with space, expanding the creative limits of dance. The experiments in technology also encouraged the viewer to question the interactivity among bodies, machines, and images, interacting reality and virtuality, pointing to technology as a powerful character in dance creation. Both productions questioned audiences about the construction of poetics from the relationship established between dance and technology, to the expansion of their perceptual field in relation to the body, movement, and dance, as well as the relations promoted by the mediation of dance by technology and how these reveal the social and political cultural aspects present in their existential reality.

References

1. Amorim, B.: Dança Contemporânea e Tecnologia Digital: novos suportes técnicos, novas configurações artísticas profissionais. In: VII REUNIÃO CIENTÍFICA ABRACE. V Reunião Científica da Abrace, São Paulo, pp. 01–04 (2009)
2. Angeli, D.: A Arte da Videodança: olhares intermidiáticos. Autografia, Rio de Janeiro, 218p. (2020)
3. Angeli, D.: A Poética da Videodança: narrativas de manifesto na contemporaneidade. Tese (Doutorado em Artes da Cena) – Instituto de Artes, UNICAMP. São Paulo (2022)
4. Caldas, P. (org.). Dança em foco: ensaios contemporâneos de videodança. Tradução de Ricardo Quintana. Aeroplano, Rio de Janeiro, RJ, 345 p. (2012)
5. Carvalho, T.M.S., Pronsato, L.: Interações entre dança e tecnologia: um estudo prático-teórico sobre a dança mediada por dispositivos tecnológicos. Palíndromo, Florianópolis **12**(26), 110–124 (2020)
6. Covaleski, R.L.: Artes e comunicação: a construção de imagens e imaginários híbridos. Galáxia **24**, 89–101 (2012)
7. Levy, P.: O que é o virtual? Editora, São Paulo 34 (2001)
8. Peirce, C.S.: Semiótica. Tradução de José Teixeira Coelho Neto. 3. ed. Perspectiva, São Paulo, SP (2005)
9. Santana, I.: Dança na cultura digital [online]. EDUFBA, Salvador, 204 p. (2006)
10. Santos, M.M.: Do teu olho sou o olhar: sobre intenções, mediações e diálogos no cinema. BOCC (2013)

Computational Art and the Creative Process

Art as an Expanded Field: The Case of the R/Place Social Experiment

Marcela Jatene Cavalcante Botelho[(⊠)] [iD] and Hosana Celeste Oliveira[(⊠)] [iD]

Universidade Federal do Pará, Belém, PA 66075-110, Brazil
marcela.botelho@ica.ufpa.br, hosana.celeste@googlemail.com

Abstract. In this paper, we will present the social experiment originating from the website Reddit, r/place, and our findings regarding how this experiment has been analyzed and understood, with three interpretations of our studies that could be associated with r/place. On top of that, we will introduce our art research methodology, "art as an expanded field," in which r/place will be our subject of analysis, with the concept of artification described by scholar Ellen Dissanayake as our theoretical starting point for such an endeavor. It is not in the interest of this work to thoroughly analyze r/place but to present a paradigm for understanding it and possibly further experiments, as well as present the beginnings of our methodology currently developed.

Keywords: R/Place · Artification · Art as an Expanded Field

1 Introduction

For the most assiduous internet users, Reddit, the website known as "the front page of the internet" [1], is a constant presence. Created in 2005 by two North American college students, the forum-like site has more than 50 million users (also known as "redditors") that access it daily [2]. Reddit is not uniform by any means, having thousands of subcommunities, called subreddits, created and managed by its users, ranging from fans of a movie series, celebrities, music bands, video games, and even the citizens of a country. Content is also mainly created and curated by redditors, and while it started as an aggregation website of what people found interesting to share at the time, Reddit has evolved to become a self-referential community with its brand of humor and jokes [1].

Introducing Reddit makes itself necessary for setting the stage of our object of study, a social experiment created by the website and executed by its users: r/place. While the Place experiment itself is multifaceted and open to many study approaches, we opt to attempt to analyze it from an art perspective that includes theories such as pixel art, net art, interactive art and artification - a many-sided method which will further on be described as an "expanded field". We also will focus on certain specific aspects of r/place: its "metamorphosis" over the 72 h of the experiment, as well the isolated drawings and composing symbols of the canvas identified by a user-created project called "r/place Atlas".

A. L. Brooks (Ed.): ArtsIT 2023, LNICST 564, pp. 123–134, 2024.
https://doi.org/10.1007/978-3-031-55319-6_9

2　When It All Started

Starting as an April Fool's joke, r/place (named as such by the way subreddits are coined in Reddit, with the "r/" prefix) was created as a social experiment in 2017 by software engineer Josh Wardle, at the time working for Reddit, and repeated in 2022. At the time of writing, r/place had its third edition which elapsed from July 20 to July 25. The idea was as follows: a blank 1000 × 1000 pixel empty canvas in which, over 72 h uninterruptedly, Reddit-registered users could change a single pixel color (or "place a tile", as put by Wardle himself) at a 5-min interval, out of a 16-color palette. However, during this 5-min gap, another number of users could overwrite the same tile, highlighting the weight of Wardle's words and the only true "instructions" of r/place: "Individually you can create something. Together you can create something more" (Fig. 1).

Fig. 1. Josh Wardle's post with the introduction and instructions for r/place.

Even if a user stayed online during the whole 72 h of r/place, placing tiles precisely at their 5-min interval, they would only be able to place a total of 864 tiles, which equates to not even 0,1% of the total area of 1 million pixels [3]. The insignificance of solitary actions thus highlighted the importance and potential of collaboration.

Still, during the first 24 h what happened was exactly what one would expect of unrestrained actions in an online environment: chaos. The canvas had no sort of visual cohesiveness, with only individual or very small groups of users placing tiles, to initially test the system or create personal works [4]. As pointed out by Wardle himself, the start of r/place looked like something out of a bathroom stall: nazi symbols and penis drawings [5] (Fig. 2).

Yet, despite initial expectations of those following the experiment, participants were able to self-organize in a particular manner [6] to create a canvas that represented what was culturally relevant for the online community at that time, with elements such as memes, jokes, and figures depicting Reddit's subculture.

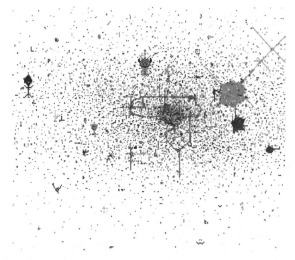

Fig. 2. The first 2 h of the r/place canvas marked by spaced-out unitary pixels, crude drawings, and phallic images.

2.1 Self-organization

What motivated this so-called "self-organization" in the first place, seeing as very disorganized actions marked the first day of the event? The growing absence of "free tiles", for one: as the number of participants increased, gradually the untouched tiles were disappearing, forcing individuals to either lose their tiles or engage in a collaborative effort to defend their territory, something that would be impossible to do alone considering the technical aspects of r/place. And the other, equally important factor, was simply the desire from Reddit's online communities, the subreddits, to leave their mark: using the website's own platform, as well as other means of communication such as Skype, Discord and Twitter, users from a same subreddit could work in an organized fashion so as to make a cohesive visual representation of their community (Fig. 3).

Besides the pre-existing communities, r/place was the birthplace of its own new subreddits and groups that were formed during the event. Worth citing here are two examples: r/TheVoid, a group of users with self-proclaimed nihilistic values that aimed to fill the canvas only with black tiles, and r/RainbowRoad, inspired by the traditional race track from the video game series Mario Kart, tried to spread rainbows diagonally across the canvas (Fig. 4).

As previously established, r/place was not static. In fact, every second, or millisecond, the canvas was suffering changes and transforming itself, with entire areas disappearing or reappearing. Because of that, what is considered the result of r/place, or "the last frame" as it is known by the users, is available at a user-created project known as r/place Atlas [7], a website that displays what is considered by many as the "final" form of r/place.

The r/place Atlas isolates each individually identifiable image from the canvas's final form, allowing users to interact with them by hovering their mouse cursor and displaying further information about that image and the smaller symbols that compose them, such

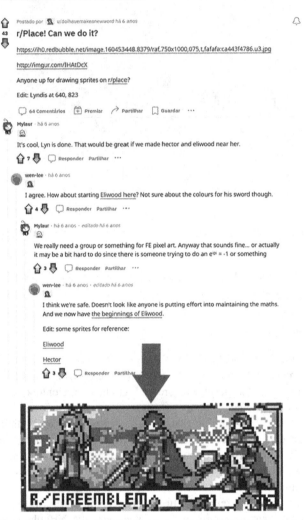

Fig. 3. A post in the video game series *Fire Emblem* subreddit inviting its members to participate in r/place. Below is the finished artwork made by them after 72 h.

as which community was responsible for it, the "attacks" it might have suffered from other communities, how it changed and more. Seeing as more than one million people participated, as well as thousands of subreddits, one that observes the canvas at first glance might not recognize the meaning and cultural relevance of every visual aspect of r/place, so the Atlas works, as its name implies, to bring this knowledge to others (Fig. 5).

By looking at the last frame, it's noticeable how the size of the images in the canvas is directly related to the size of the community it represents. Meaning that subreddits of countries or soccer teams, for example, have proportionally bigger symbols than those of smaller subreddits, such as niche books or video game series. Concerning this, one of the

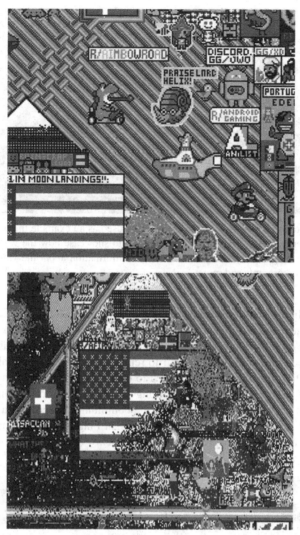

Fig. 4. At the top of the image a depiction of r/RainbowRoad, with their signature and rainbows growing across the canvas. Below is an example of an "attack" by r/TheVoid trying to cover the canvas in black pixels.

most peculiar aspects of r/place was precisely the behavior exhibited by the communities participating: because it was a global event, users had to deal with differences in time zones, meaning that a time period where the majority of members from a subreddit would be asleep or at work implies that their image, or "territory" would be susceptible to vandalizing or even being overtaken by other subreddits with a different time zone. This problem led to many alliances, with some allies forming their images next to one another to facilitate protection at every possible hour.

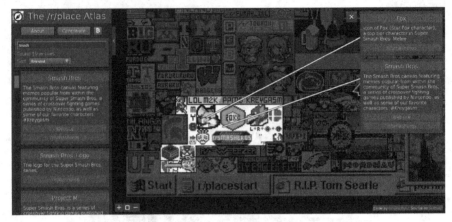

Fig. 5. The r/place Atlas interface. In this picture, we used as an example the image created by the videogame series *Super Smash Bros.* Community to showcase how the website works. By hovering over the bigger picture, the website reveals information about the video game series, what each symbol represents and more.

Not only alliances but rivalries were born – for many reasons – and registered both in the Atlas project and as timelapse videos available on Reddit and other video hosting sites, such as YouTube. One of the most curious rival interactions was between the subreddits of Germany and France. As if to replicate history, intentionally or not, the image representing the German flag started to overtake the image of the French flag, reminiscing the nazi invasion of France during the Second World War on June 22nd of 1940. As to not end on such a tragic note, the communities seem to have reached an agreement, and the space conquered by the German Flag eventually transformed into the European Union flag, as depicted in the figure below (Fig. 6).

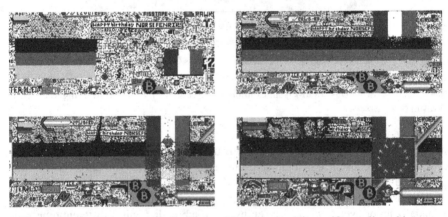

Fig. 6. The rivalry between the German and French subreddits, as if to replicate history.

Seeing as the total scope of the canvas is 1 million pixels, this is but one of the many tales it has to offer. Considering the restrictions of this paper, it would be impossible to talk about each of them, and we can argue that such varied happenings surrounding this social experiment is but a testament to its magnitude and its narrative potential. For this reason, as well as many others, r/place is studied and analyzed by many different disciplinary fields and approaches, which we'll cover in the following section.

2.2 A Social Experiment with Artification Features

How to define r/place? We have done an extensive literature review to determine how r/place is defined and which terms are used. Our findings are presented here by category: in the Social Sciences, r/place is as a game, experiment, or competition [3, 8, 9]; in the field of Information Technology, it goes by names such as sandbox, collaborative project, collaborative social experiment, mass controlled event, online experiment, and peer-production platform [10–13]; in the Arts (Visual Arts, Theater, Music and Design) understand r/place as a mosaic art [9], crowdsourced art [14] and the action of its users as artistic endeavors [10]. It is relevant to notice that depending on the field, the terms vary, yet they are not mutually exclusive, which shows that r/place is approachable from different perspectives. At this point of our research, we have chosen to refer to r/place as a "social experiment with artification features".

Specifically in the literature that focuses on r/place under the optics of the Arts, authors understand it either in its complete form [15–17], isolated images and symbols [4] such as those identified by the Atlas project, yet none of these works defines r/place as a specific type of Art related either to a determined style or stylistic movement. Despite focusing on Art - specifically Pixel art [18–21], Net Art [22–26], and Interactive Art [27] - we consider other fields as complementary to our investigation on r/place since only one point of view would leave many blanks to consider, and to solve this problem, is necessary a multi/inter/transdisciplinary approach.

As Pixel Art

One possible approach to r/place is Pixel Art. "Pixel" is an abbreviation of the expression "picture element", first mentioned in 1965 on image processing and analysis [20]. In digital images, the pixel is considered the smallest component, not to say that all pixels are the same size or even the same format: its size varies from screen to screen. On the other hand, Pixel Art first appeared in work published for the Association for Computing Machinery (ACM) [19] in 1982, referring to an image that was scanned and then digitally modified. In the 80s, personal computers with colored displays were becoming more common among the general populace, and the artists of the time started experimenting with the possibilities of pixel art [18].

Nevertheless, if pixels constitutes every digital image, could we say that, as such, every single digital image is Pixel Art? Not reasonably, since pixel art is described as having every pixel visible and intentionally placed [21]. For example, vector images, created by a different method, despite being digital and composed of pixels, are made by manipulating points, lines, and curves, not pixels directly, and low-resolution images might have their pixels "visible." However, they were not intentionally placed but rather just a result of image processing.

Established these criteria, could r/place be considered pixel art? At first glance, being such a large-scale canvas, one could argue that the pixels are not immediately visible, but that is a matter of size and perspective. The pixels are also intentionally placed, despite being named "tiles". So yes, we could consider r/place as an example of pixel art, as its entire canvas as a single unit of art or as the canvas being used as an exhibition space composed of other more minor, size-varying Pixel Art.

As Net Art

We could also look at r/place from the perspective of Net Art. The works considered part of the Net Art movement challenged notions of authorship and aesthetics and acknowledged the process, not only the finished piece, as part of the work's identity.

These aspects are easily applied to r/place, a social experiment without indeed an "author" responsible for it, having over 1 million users participating, has many different aesthetics, and the changes it went through in the duration of the experiment are as equally important to its meaning as it is the last frame.

It is also essential to consider that when referring to "Net Art," we are also considering many of the other terms and concepts used by many different authors that relate roughly to the same ideas, including Web Art, Digital Art, Internet Engaged Art, Internet Aware-Art and Post-Internet Art [22–26, 28, 29]. Generally, these terms are used interchangeably, even though a more incisive study might point towards some and other peculiarities. Even so, this is not the main topic of this paper and will not be discussed further.

As Interactive Art

The approach of Interactive Art is also a possibility to understand r/place. One research focused on Human-Computer Interaction (HCI) classifies r/place as Interactive Art, in which the public and the art piece form an interactive dialogue with the potential of being unique to each participant [30]. This specific work establishes four types of interactive art: virtual, embodied, tangible, and social, with r/place falling under the last one, social, being capable of encouraging people to collaborate with others and reach complex levels of self-organization.

Although this research belongs to Computer Science, the authors strongly rely on theories from interactive artworks [27]. However, we chose not to include the terms used to define r/place in our literature review above since the work itself is not from the field of Arts.

Which of the above could better understand r/place – Pixel Art, Net Art, or Interactive Art – is not a question we hope to answer with this paper. By presenting these three possible comprehensions under the scope of the Arts, we do not expect to define which is the most well-suited for studying r/place, but rather as possibilities to develop methodologies that can look at r/place as a social experiment with artification features and assume the importance of a broader scope of theories to analyze it.

3   Methodological Proposal to Study R/Place: Art as an Expanded Field + Artification Behavior

To be better equipped to analyze r/place and its branching characteristics, we propose our methodology therefore known as "expanded field." It is also important to clarify that this methodology is in the process of being developed, so what we aim to present here is but an introduction to its core aspects and, briefly, one of the primary theories that will serve as a basis for this framework of analysis and study. It is not our intention at this time to be applying the methodology in its total capacity to analyze r/place, but to set ground for further developments in our research regarding it and other subjects that we encounter and see to experiment on with this methodology.

Art as an expanded field can be understood as an approach to studying and comprehending art in its many forms from an inter/multi/transdisciplinary perspective embracing different types of knowledge, without undermining the peculiar characteristics that art itself has as a creative endeavor or field of study, and being able to associate art, with the characteristics just described, with varied realms of research such as science, technology, among others. In this first experiment with the methodology, we embrace the concept of "artification", as described by Ellen Dissanayake [31] in her many works on the subject, as the first step towards the expanded field of art.

Dissanayake describes artification as the noun of the verb "to artify", that is, "the capacity to make the ordinary everyday experience 'extraordinary,' or special," an ability exclusive to human beings and associated with its evolution over the millennia. Artification is, as such, part of the "making" and not the "result," turning our attention towards the process of art and its result rather than the finished piece itself. It should not be mistaken for "artmaking" as we know it formally, and instead something else: in Dissanayake words, "artification, unlike 'art', may be unskilled, unoriginal and even pedestrian" - the crux lies on the action, the process, and the "extraordinary" out of the ordinary.

We can relate the idea of artification to the r/place experiment: rather than focusing on the finished canvas alone, we observe the process of it, of how millions of users take an ordinary white canvas and transform (or *artify*) it to make it unique for them. Many of the images constructed are, in fact, copies of other artworks: the Mona Lisa, national flags, panels from comic books and *manga*, which one could see as unoriginal or even pedestrian, yet, the focus shifts from their standalone forms, and consider the behavior towards a collective expression, of *making an area of the canvas unique* to them and can see how artification easily meshes with r/place, and how it can be a valuable tool for analyzing it.

Although, as previously established, we do not aim to offer one solution to understanding r/place, and artification is not the only possible way, we present it as a novel way of seeing it and hope to develop the expanded field further to encompass many more fields and theories (Fig. 7).

Fig. 7. The last frame of r/place.

4 Conclusions

R/place has yet to end in its last frame, not even in this work. It is a social experiment, collaborative art form, capable of being studied from the perspective of Pixel Art, Net Art, and Interactive Art, as well as from a multi/inter/transdisciplinary point of view that can help reignite studies in the Arts field that interact with science and technology, as well as social sciences, in equal standing. Specifically in the field of Arts, r/place can help dissolve barriers imposed by styles, terminologies, and formal classifications, helping occupy new spaces that consider the process as part of the artwork as well as the entire experience of it.

Moving forwards, our research on r/place is still ongoing. We hope to expand the ways we can analyze it, developing our methodology based on the principles mentioned above, and that it can also be used to comprehend other similar works around the internet, without, in any way, the presumption of being the only applicable methodology for such. To expand beyond online collaborative works and accept all sorts of experiences and processes under the arts, science, and social studies to further what is "to do art."

Acknowledgements. This research is funded by CAPES through a M.A. scholarship, and by UFPA through conference funding.

References

1. Singer, P., Flöck, F., Meinhart, C., Zeitfogel, E., Strohmaier, M.: Evolution of reddit: from the front page of the internet to a self-referential community? In: Proceedings of the 23rd International Conference on World Wide Web, Seoul Korea, pp. 517–522. ACM (2014). https://doi.org/10.1145/2567948.2576943
2. Patel, S.: Reddit Claims 52 Million Daily Users, Revealing a Key Figure for Social-Media Platforms. https://www.wsj.com/articles/reddit-claims-52-million-daily-users-reveal ing-a-key-figure-for-social-media-platforms-11606822200
3. Lauria, P.A.B.: As dinâmicas provenientes da espacialização, competição e cooperação dos usuários e comunidades virtuais do Reddit (2019). https://dl.acm.org/doi/10.1145/3274192. 3274227
4. Gürkan, A.: Place! Steal! Design! The Use of Game in the Urban Design Practices (2021). https://tase21.artun.ee/wp-content/uploads/2021/06/Artun-Gurkan_Thesis.pdf
5. Cuthbertson, A.: Reddit Place: The Internet's Best Experiment Yet. https://www.newsweek. com/reddit-place-internet-experiment-579049
6. Adams, A.M., Fernandez, J., Witkowski, O.: Two Ways of Understanding Social Dynamics: Analyzing the Predictability of Emergence of Objects in Reddit r/place Dependent on Locality in Space and Time (2022)
7. The /r/place Atlas. https://draemm.li/various/place-atlas/. Accessed 09 Aug 2023
8. Mørch, A.I., Andersen, R., Kaliisa, R., Litherland, K.: Mixed methods with social network analysis for networked learning: lessons learned from three case studies. In: Networked Learning Conference, Norway (2020)
9. Ganguly, S.: Declining Entropy on Online Forums - A Contrarian Case Study. Declin. Entropy Online Forums - Contrar. Case Study (2022)
10. Armstrong, B.: Coordination in a Peer Production Platform: A study of Reddit's /r/Place experiment (2018)
11. Aleta, A., Moreno, Y.: The dynamics of collective social behavior in a crowd controlled game. EPJ Data Sci. **8**, 22 (2019). https://doi.org/10.1140/epjds/s13688-019-0200-1
12. Lapid, S., Kagan, D., Fire, M.: Co-membership-based generic anomalous communities detection. Neural. Process. Lett. (2023). https://doi.org/10.1007/s11063-022-11103-1
13. Rappaz, J.: Dynamic Personalized Ranking (2023). https://infoscience.epfl.ch/record/296038/ files/EPFL_TH7978.pdf
14. Palonis, B.: Understanding the Themes of Disability Discourse Through Reddit Comments (2021)
15. Gimeno, R.A.: Creación de un juego colaborativo masivo en línea (2018). https://diposit.ub. edu/dspace/bitstream/2445/130477/3/memoria.pdf
16. Monzó, M.G.: Redes para una Sociedad de Supervivencia. Proyecto de instalación interactiva auto-organizativa (2021). http://polipapers.upv.es/index.php/IA/article/view/3293
17. Sommeregger, E., Messini, V.: Posthumanist Sandbox: The Potential of Multiplayer – Environments, Austria (2021)
18. Clark, S., Davis, G.: Revisiting and re-presenting 1980s micro computer art. Presented at the Proceedings of EVA London 2021 (2021). https://doi.org/10.14236/ewic/EVA2021.52
19. Goldberg, A., Flegal, R.: Pixel Art. Pixel Art. **25**, 2 (1982)
20. Lyon, R.F.: A brief history of "pixel." Presented at the Electronic Imaging 2006, San Jose, CA, 2 February (2006). https://doi.org/10.1117/12.644941
21. Silber, D.: Pixel Art for Game Developers. CRC Press, Boca Raton (2016)
22. Bulhões, M.A.: Desafios: arte e internet no Brasil. Editora Zouk (2022)
23. Nunes, F.O.: Ctrl+alt+del: distúrbios em arte e tecnologia. FAPESP: Perspectiva, São Paulo (2010)

24. Paul, C. (ed.): A Companion to Digital Art. Wiley Blackwell, Malden (2016)
25. Stallabrass, J.: Can art history digest net art? Can Art Hist. Dig. Net Art. 165–179 (2009)
26. Stallabrass, J.: The Aesthetics of Net.art. Qui Parle **14**, 49–72 (2003). https://doi.org/10.1215/quiparle.14.1.49
27. Candy, L., Ferguson, S. (eds.): Interactive Experience in the Digital Age: Evaluating New Art Practice. Springer, Cham (2014). https://doi.org/10.1007/978-3-319-04510-8
28. Giaccardi, E.: Movements and passages: the legacy of net art. Mov. Passages Leg. Net Art. **13**, 23–32 (2005)
29. Quaranta, D.: Situating post internet. In: Catricalà, V. (ed.) Media Art: Towards a New Definition of Arts in the Age of Technology. Gli ori, Pistoia (2015)
30. Duarte, E.F., Baranauskas, M.C.C.: Revisiting interactive art from an interaction design perspective: opening a research agenda. In: Proceedings of the 17th Brazilian Symposium on Human Factors in Computing Systems, Belém, Brazil, pp. 1–10. ACM (2018). https://doi.org/10.1145/3274192.3274227
31. Dissanayake, E.: The concept of artification. In: Early Rock Art of the American West: The Geometric Enigma, pp. 91–129 (2018)

OPHILIA: Cy-Collage Cyberperformance

Rosimária Sapucaia[1,2](✉) ⓘ, Célia Vieira[1,2] ⓘ, Inês Guerra Santos[1,2] ⓘ,
Ana Carvalho[1,2], and Juliana Wexel[3] ⓘ

[1] University of Maia, Maia, Portugal
`{cvieira,iguerra,anamariacarvalho}@umaia.pt`
[2] CITEI/CIAC, University of Algarve, Porto, Portugal
`rsrocha@ualg.pt`
[3] University of Algarve - Open University/CIAC, Sicily, Italy

Abstract. This communication intends to present a practical case developed within the context of an exploratory project on cyberperformance, titled CyPeT. As part of this project, we surveyed the strategies developed by teachers, creators and performers during periods of confinement and identified the possibilities and challenges that cyberperformance brought to the context of the performing arts. Resulting from a collaborative creative process, the experimental cyberperformance *Ophilia*, created by the team involved in this study, aims at learning more about creation in a digital medium through practice. *Ophilia* is a work in progress, developed through the implementation of an arts-based research methodology, that allowed the team of researchers/creators to identify creative strategies inherent to cyberperformance while adapting Shakespeare's character Ophelia to a new medium. Above all, during this process, we were able to identify various technological limitations inherent to the platforms provided by educational institutions when applied to the creation of cyberperformances.

Keywords: Cyberperformance · Artistic practices · Teaching methodologies · Higher Education · arts-based methodology

1 The CyPeT Project and the Cy Co-lab Pedagogical Experimentation Workshop

CyPeT is a research project funded by the Foundation for Science and Technology (FCT) in Portugal started in 2021. The aim of the project was to seek critical thinking about performance art produced online and envisage new teaching methodologies that could arise from this investigation. To this end, we analysed the strategies developed for the teaching of performing arts during confinement and developed a diagnosis of the academic reality, during the same period, based on the analysis of the results of the surveys applied to Portuguese Higher Education Institutions (HEIs). We also analysed the artistic practices, based on the results from semi-structured interviews with artists who, during the period of confinement, used new strategies for the presentation and dissemination of their performances.

A. L. Brooks (Ed.): ArtsIT 2023, LNICST 564, pp. 135–148, 2024.
https://doi.org/10.1007/978-3-031-55319-6_10

Online teaching, imposed throughout the confinement periods, in the years 2020–2021, has faced many challenges. The "new normal", as this phase became known, demanded from teachers/artists the ability to "reinvent themselves", to adapt, and to be creative in even more precarious conditions than usual [1]. In the Portuguese context, studies on this period, in education, are still scarce. Most publications focus on how institutions have adapted to remote teaching, emphasizing technological requirements, teaching strategies, and means for interacting with students [2]. Even scarcer is the research related to online teaching within performing arts. The existing few present generalist studies: the authors [1, 3] and [4] point to online teaching during confinement while [5] and [6] address specifically research focused on online teaching in performing arts.

All the investigations mentioned above put forward a series of problems and challenges associated with the abrupt transition from face-to-face teaching to emergency remote teaching. Among the main difficulties identified by the authors are the conditions for distance learning, namely network problems, the lack of adequate equipment and the diversity (and inequality) in the access to technological resources felt by students and teachers [2]. Furthermore, in the next section, we deepen this subject in the analysis of surveys carried out with Portuguese HEIs.

Despite technological difficulties, creativity allowed good practices and successful experiences. Creativity is inherent to human beings, making possible individual discoveries and reinventing oneself in different circumstances. It is present in our society in various instances, enabling scientific, cultural, political, economic and educational advances over time. In the university context, it could not be different, as creative processes are important for structuring and thinking about didactic structures for the classroom. As such, "it is observed that the world has changed quickly, however, many didactic formats used in the classroom remain the same" [7]. The emergence of distance communications became a pressing issue in Higher Education that demanded creative solutions: how to adapt the teaching of artistic practices that traditionally require corporeality to a virtual space that dematerializes it? How can creative adaptation be carried out by teachers without previous technological training? These questions became even more relevant because the pandemic situation highlighted the current gap between, on the one hand, the manifestations of a society that tends to disengage from traditional cultural media (theatre, cinema, books) preferring digital content, and, on the other hand, the absence of effective methodologies for teaching in the arts context through digital devices.

When considering the transposition of the classroom from face-to-face to online environments [8], is necessary to go beyond the physical space, going also beyond the structure of the classroom and the traditional way of thinking about it. Society works as a diverse and complex network and creative education has the challenge of keeping up with this dynamism, by including collaborative practices to boosts student's creativity, projects, and creative processes. In this sense, the study of cyberperformance responds to the developments of innovating creatively in the teaching methodologies for online performing arts.

Throughout the research and in addition to presenting the state of the art on cyberperformance, questionnaires were carried out on higher education institutions and interviews with artists. This constituted the first stage of the CyPeT project. On the second stage, for the development of a methodology aimed at teaching cyberperformance, artistic works were created. One of these projects, Ophilia, was developed at a workshop on pedagogical experimentation, which took place in May 2023, titled Cy Co-lab. The Cy Co-lab workshop brought together teachers and artists for a one-day event that aimed at contributing to the development of a pedagogical model to be implemented within the scope of the CyPeT project. The event has as objectives: to share knowledge, perspectives and strategies of artistic and pedagogical nature in the area of cyberperformance by involving participants from different areas: teachers, artists and technicians; to propose adjustments to a pedagogical model for teaching cyberperformance under development within the CyPeT project; the creation of microcyberperformances; to develop critical thinking on the process of creating a cyberperformance, along with the teaching-learning process of a digital performance.

2 Cyberperformance

The term Cyberperformance describes live performance that takes place on digital platforms, and it is characterized by being mediated, intermedial, hybrid, collaborative, aesthetically and socially intervening [9–11]. Cyberperformance adapts tools, questions the cartesian division between mind and body, and the concept of simulating reality. It demonstrates that virtual reality is as real and consequential as the physical world. Its genealogy combines performance and net art because the Internet is its native environment [10]. That is, Cyberperformance draws on several sources, but it is mainly dependent on the Internet, reason why it tends to be processual and open ended [10, 12, 13]. As an artistic expression, cyberperformance is also described by new media characteristics, such as connectivity, universality, dissemination speed and virtuality, which distinguish these from other media-based artistic practices and establish unique possibilities that lead to new creative processes [14].

Attending to its characteristics, cyberperformance has become a departing practice for the development of creative methodologies in the online teaching, precisely because of the it establishes connections between the existing online dynamics through digital platforms and social networks as privileged spaces for the development of a "pedagogy of connections" [15]. From here, new initiatives gradually emerge that evoke engagement and personalization through the articulation of contents and skills of the digital and mobility era, by acknowledging the potential of the network communities and new concepts of participation, collaboration, autonomous learning, and associated multidirectional and creative dynamics.

As consequence, there are necessarily distinctive elements of this medium [16] which, as our study has shown [17], can condition the creation, dissemination, and reception of the work. The aesthetics of the Internet imply an immersion effect. Surely, immersion can be provoked by any other medium (music, literature, etc.), any time the listener/reader/spectator experiences the effect of being immersed in another reality in which all their senses are absorbed and become oblivious to the reality in which they find themselves. But the Internet has additional features that enhance this effect, for example, through the possibility of virtually entering an augmented reality, and through the multisensory stimuli that the devices allow (including video, sound, photography, graphic productions, animation, etc.).

Furthermore, immersion in virtual environments is altered by participation. Participation, enhanced by interaction, is not exactly original to the Internet, as it finds equivalence, for example, in theatre, with the fall of the "fourth wall". The difference, though, lies in the fact that participation online is inherent to the medium, part of the cybernetic creative process that totally obliterates the performative geographical distance. Participants must register, they have to create an access name, they can comment on the performance, synchronously or not, in short, there is the possibility of interaction between the participants/visitors and the performers. Another characteristic of this medium is its rhizomatic essence, the navigability resulting from its inherent transtextuality. Unlike a theatre, where the spectator is invited to disconnect, on a virtual stage the visitor/participant has countless tabs, toolbars, links, etc. at their disposal. One of the most frequent criticisms of the audience's behavior in online events concerns the fact that viewers turn off the camera (often to perform other actions at the same time), especially when the viewer realizes that they must participate actively, they have to play the role of performer and not just be a passive spectator. In this sense, the Internet imposes a redefinition of the spectator's role since it demands taking an attitude and deciding to be present or to disconnect.

Among the many features in the definition of the Internet as a medium for the performing arts and, consequently, of the dramatic productions it generates, we would like to highlight two as the most relevant taking in consideration the results of current research. One feature concerns the fact that the Internet allows the convergence, in a virtual space, of voices, texts, images, sounds, locations, which may be distant in time and space both real and fictional. The other feature concerns the error, failure and noise that can occur at any time because of the instability of the medium and its mediated nature and the way in which this randomness integrates the ephemeral aesthetics of the work created.

3 Methodology

3.1 Collecting Data for a Diagnosis

As already mentioned, we found a need to make a diagnosis that would address the difficulties and strategies found by teachers to mitigate or to overcome the constraints resulting from the pandemic. To this end, all the coordinators of the first cycles of study in performing arts in Portuguese Higher Education Institutions (HEIs) were invited to participate on the survey "Online Teaching of the performing arts in a context of confinement", of which twenty-three participated in a study caried out between April and July 2022.

To what concerns the semi-structured interviews with artists, the corpus was defined by an intentional sample of artists and artistic groups. Six interviews were made.

While the results of the questionnaires were subject to statistical treatment, the interviews were subject to content analysis according to a qualitative model [18].

The survey includes open and closed questions related to cyberperformance, creativity and Higher Education with the intention of collecting data about the impact of the pandemic on teaching performance related classes.

Before carrying out the interviews, guidelines organized by topics were created which included questions referring to the challenges that COVID-19 brought to creativity and the strategies that Higher Education Institutions and their agents have designed to overcome the challenges. The participants also detailed what they did during the confinement and what they think could continue afterwards.

3.2 Arts-Based Research

Arts-based research is a transdisciplinary strategy that values the role of creative arts in research contexts [19]. One of its main characteristics, and simultaneously one of the major potentialities of this methodology, is the fact that it favors interaction (at different levels: both emotional and sensorial) between the researcher/artist and the audience. As such, the process becomes particularly important, sometimes even more than the final outcome. This methodology involves a critical, and simultaneously collaborative, reflective attitude of everyone involved.

Artistic practices are at the core of the research itself [20]. This does not mean, however, that all research is limited to the practice. In fact, in many cases this methodology is combined with the use of other methodologies of an empirical nature, such as interviews or mixed with quantitative analysis.

Arts-based research expands the limits of discursive communication to artistic practices, where transmitting meanings and sensations is hardly apparent when using traditional methodologies. It is precisely why methodologies based on the practice of researchers/artists can lead to directions that are impossible to achieve with traditional scientific methodologies [21].

Research that is based on or as artistic practice is not just research "about art", but rather "through art" [22]. The creative process within the digital arts is traversed horizontally through long cycles of deep reflection. Artists continually raise questions about primordial visions and initial concepts/ideas under the light of the "other", being this other the audience, the performer, or even the materials, the technologies, the tools, and the prototype of the artifacts. But the arts-based methodology is not exclusively centered on the individual, rather, it is based on collaborative logics where the creative process and the research product(s) are shared. Precisely for this reason, it has been demonstrated that its effect tends to be more effective and immediate, particularly when applied to an educational context.

Artistic practices have proven to be particularly fertile in generating new curriculum approaches for general educational practices and the development of the learners [21].

The cyberperformance Ophilia and its entire creative process is the result of arts-based methodology studied and developed throughout the project CyPeT.

4 *Ophilia*: A Work in Process

The cyberperformance Ophilia (Fig. 1) is one of the microcyberperformances developed as part of the Cy Co-lab pedagogical experimentation workshop, in May 2023 (Fig. 2). After the workshop, collectively, the participants decided to present the performance at the Cyberperformance Symposium, the final event of the CyPeT project, on the 29 and 30 of June 2023, at the University of Algarve.

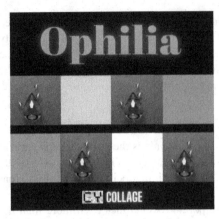

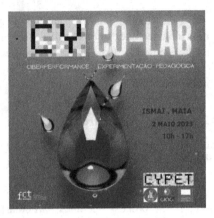

Fig. 1. Poster of the presentation of the cyberperformance Ophilia

Fig. 2. The poster of the Cy Co-lab pedagogical experimentation workshop.

Together with the authors of this text, took part on Cy Co-lab many invited artists and teachers from the group of respondents to the interviews. This was the case with Pedro Alves, artistic director of Teatromosca (the second participant from the right, Fig. 3) and with two other professors from Unicamp, Brazil, who developed the cyberperformative Jardim das Cartas. The researchers are: the scenic artist and researcher at the Institute of Arts, Department of Performing Arts, Daniela Gatti (the third participant from the right, Fig. 3), and composer, researcher, and teacher at the Institute of Arts, Jônatas Manzolli (the first participant on Fig. 3). Manzolli is also a Visiting Full Professor at CISUC-DEI, University of Coimbra, and at DeCa/University of Aveiro. Cy Co-lab took place on Microsoft Teams platform and at University of Maia campus and was divided into two moments. The first moment, held in the morning, in addition to presentations by the above-mentioned guests, the authors of this article shared CyPeT's results, the state of the art, and case studies. On the second part, Jônatas Manzolli presented a few applications used to create the cyberperformance Jardim das Cartas. Collectively, all the participants engaged on a creative process from which resulted the cyberperformance Ophilia.

The ephemeral collective at the Cy Co-lab chose to use the principle of *collage* as an aesthetic technique to combine various textual, sound and image elements. Like *montage* and *assemblage*, *collage* is a central technique of the modern visual art and fundamental to artistic movements such as Cubism [22]. It allows materials of different origins to be combined, providing a unifying meaning to the composition. In the case of Ophilia, collage has been adapted to intersect modern art and the performing arts through digital technologies [23].

Water was chosen as the central theme for Ophilia in connection with the concepts of fluidity and integration. The theme seeks to present how collectively, teachers and artists can engage in dialogue to overcome obstacles and learn with each other. Water is a great metaphor for our work together as it relates with fluidity, mutability, and the way in which, in constant movement, different particles can be united, allowing drift while leading to converge into new forms. The theme also related closely to the

Fig. 3. Workshop Cy Co-lab on May 2, 2023. Photographer Juliana Wexel (2023).

textual tradition of Shakespeare's Ophelia which led to new sub-themes associated: the theme of love and the theme of woman. The tragic character of Ophelia was adapted to become multiple cybernetic Ophelias. In digital contexts, new forms of adaptation have been developed by establishing connections between media through dynamics of transmediation. These processes led to developments in the theory of adaptation providing perspectives that allow traditional comparative approaches to be rethought. In literature and film studies, for example, one of the dominant trends has been to focus on the question of transposition i.e., on analysis based on the comparison of works resulting from the transition from one medium to another, more precisely, from the literary form to the moving image, systematising similarities, and differences. The question of adaptation raises a broader issue, that of the function of adaptation in digital culture. Multimedia adaptation highlights creative processes in which adaptation is not simply rewriting a concrete, finished work, but rather integrates forms of creation that tend to be transtextual and transmedia. This perspective corresponds to a notion of adaptation as fluid, in the sense of Bryant [24], which includes all forms of transferring, rewriting, and revisiting a text, in other words, a notion of adaptation as an "ever-changing ever-developing story" [25]. From Shakespeare's tragic Ophelia to the creation of the cyberperformance Ophilia, there is not only a different thematic approach, instead, several transformations take place resulting from the change of the medium itself [16].

Using adaptation as a creative strategy, new and contemporary versions of the tragic Ophelia came to existence. These new modulations manifested through poetic declarations made by emancipated Ophelias, who refuse to die for love, who accuse any Hamlet of cowardice and find in water the natural element that reconnects them with cosmic harmony, far beyond any romantic disaster. The first composition, "Archimedes' Principle", by Célia Vieira, takes the form of a past hypothesis and questions what Ophelia's behaviour would be if she had understood her existence as part of the liquidity that unites all beings: "if Ophelia on her way to the river/would have found a friend/a shelter, a bird, or a book/and her blind eyes stopped/and if Ophelia had seen on her way to the river/that living or dead/the inert body floating/ […] and laughed at her image/would she continue to float/on the surface of the water?".

One other of the several poetic compositions, the poem *Ophelina*, by Juliana Wexel, breaks with the premise that water is the material of female death in literature [26]. In the first verse, the text's interlocutor, the Greek naiad Daphne, tries to dissuade Ophilia from committing the suicide of passion: "*Ah, Ophilia, Ophelina… Thy name is not frailty*". The sentence subverts Hamlet's original speech in dialogue with the Denmark's Queen Gertrude, his mother, in Shakespeare's original, "*Frailty, thy name is woman*" [27]. In the last lines, Daphne suggests that the tragic character should enjoy herself rather than end her life because of an amorous disappointment: "*Ah, Ophilia, you would sip a port wine, slowly, slowly… You would contemplate the prairies of urban madness, smile in soliloquy and say 'no, no, Laertes… That madman? He's not worth loving!'*". The text was regionalized in reference to the fundamental product of the cultural landscape of the Douro Wine Region, the oldest controlled vineyards in the world. The poem *Ophelina* has been presented at the performance *O Baile da Ofélia* (2014), produced by Denise Andrade and developed in collaboration with the artistic producer Riko Viana. The textual creation is derived from the adaptation of *Hamlet* by Vitor Pordeus, for the play *Loucura sim, mas tem seu método*, a project of the *Hotel e Spa da Loucura* (2012 and 2016), developed at the Instituto Municipal Nise da Silveira, Rio de Janeiro, Brazil.

We assumed that these lyrical texts would form modules of a multimedia composition together with spoken word, sung word, written word, video, and music. For its orchestration, we adopted a rondo structure since its ternary structure could serve as a score for the various modules. Singing worked as the connecting thread between modules.

The three movements that constitute the structure are: i) poetic dialogues between the Ophelias (inspired by the character from William Shakespeare's Hamlet), having as background a video of a river (Fig. 4); ii) interaction between Ophelia and ChatGPT, in an attempt to understand about life and love from an outside perspective (Fig. 5); iii) a poetic statement of Ophelia presented in voice-over, using a camera filter (Fig. 6).

In the dramaturgical composition of *Ophilia*, the use of *OpenAI's Web ChatGPT* interface had two main purposes. Firstly, as a digital tool, we could experiment textual construction in dialog with a cyber Ophilia and, secondly, the ChatGPT played the role of a supporting character due to its ability to imitate human conversationthrough a verisimilar interaction. This last aspect reconfigured the dramaturgy by including aspects of Artificial Intelligence Art (AI Art) in the cyberperformance.

As cyber Ophilia, Juliana begins the dialog with ChatGPT as if the virtual assistant was human: "*Dear Chat, what is love?*". She follows aloud the chatbot's superficial digression on the subject. From here, Ophilia deepens the bond of affection with the virtual assistant taking inspiration by what Bachelard calls the *Ophelia complex* and the literary preparation for suicide [26]. The conversation also references the dialogue between Ophelia and Hamlet in Shakespeare's original text: "*My 'ex' said he never loved me and told me to go to a convent… I'm thinking of diving into the waters of eternity…*". The chatbot recognizes the metaphor of suicide in the passage "*diving into the waters of eternity*" and dissuades Ophilia from her suicidal intentions by advising her to seek a specialist help. At this point, cyber Ophilia becomes even more intimate with the chatbot by comparing it to her "ex" with compliments and inviting to an in person meeting. The digital interlocutor kindly replies that this would not be possible, because, as an artificial intelligence assistant, it has no emotions and no human form. Ophilia insists and explains: *"But I like your artificial intelligence!"* Disappointed by the virtual assistant's second rejection of her invitation, repeating the same information as the previous answer, Ophilia concludes: *"I'm going to dive..."* and interrupts the conversation. The general intent of the text was based on the concept of liquid love [28] by addressing disillusionment *(pathos)* in contemporary use of dating apps.

Throughout the three moments, the singer Rosimária Sapucaia sings live the It's Sweet to Die in Love inspired by the song "É doce morrer no mar" (It's Sweet to Die at Sea), by Caymmi and Jorge Amado.

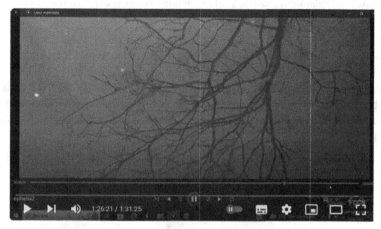

Fig. 4. Frame of *Ophilia* Performance, available at https://www.youtube.com/watch?v=5ls_Q8y MObg.

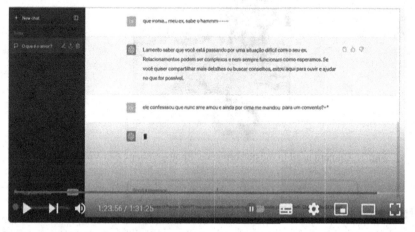

Fig.5. Frame of *Ophilia* Performance, available at https://www.youtube.com/watch?v=5ls_Q8y MObg.

The performance has a duration of 10 min. The chosen software for creating and presenting Ophilia was Zoom, unlike Cy Co-lab that used Teams. Five researchers from

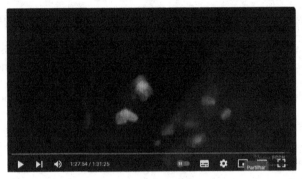

Fig.6. Frame of *Ophilia* Performance, available at https://www.youtube.com/watch?v=5ls_Q8y MObg

CyPeT were involved, from making the script to presenting the performance. The intervention of Jônatas Manzolli was also very relevant for the development of Ophilia. The equipment used was the IPVEVO VZ-R HDMI Camara, model CDVH-031P, connected to a notebook (Fig. 7). This camera helped to create the immersive environment through filters that allowed the simulation of different landscapes and effects, namely a shadow effect (Fig. 8).

Fig. 7. Camera used in the performance. Photographer Célia Vieira (2023).

5   Results

Immersion that combines physical and virtual spaces and establish collaborative networks by enhancing ways of learning and teaching that reinforce appropriation, [29]. We have considered the following steps while developing Ophilia [30]:

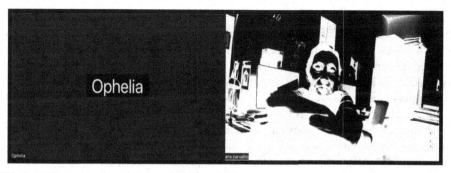

Fig. 8. Filter generated with the IPVEVO VZ-R HDMI Document Camara. Photographer Juliana Wexel (2023).

1- Network collaboration. The group gathered several times via the Zoom platform for meetings, developing the performance, experiment with glitches, effects, soundscapes, and the creation of scenarios. Throughout the process, a few elements of the group were in different countries (Italy, Brazil, Portugal) and only came together at the workshop and for the final performance. The performance in itself is not the main goal of the process, most importantly is the creative synergy generated before, during and after the performance.

2- Multi-referential space-time. The group studied and tested various platforms and artifacts. Some of these tests were done asynchronously (the videos, for example) and others synchronously (the interaction with artificial intelligence, for example). As result, spaces and digital artefacts used were the ones suitable for the cyberperformance.

3- Cybercreativity. Also associated with technological innovation, cybercreativity highlights the inventive ability to create online digital art and flexibility around technical constraints, which cease to be an impediment factor and become allies in the process of creating and displaying online performance. The group had to overcome a series of obstacles arising from the limitations found while testing the platform and accept them as part of the creation itself.

4- Autonomy and dialogue. The collaborators involved in the creative process of this cyberperformance are autonomous and weave networks of creation, training and knowledge exchange among themselves. Autonomy made it possible for each performer to be autonomous in creating their own Ophelia artistic expression. But the plural and expanded dialogue made possible the collaborative spaces-times, transforming them into self-sustainable entities and seeking a continuous status of performativity.

6   Final Considerations

Ophilia was developed by a multidisciplinary team, thorough diverse dialogues that allowed us to put into practice an entire theoretical model of artistic creation. This dialogue created possibilities for performers without experience to create while exploring the technological tools available.

During this process we were able to identify, through practice, various technological limitations inherent to the platforms provided by educational institutions regardless of if applied to the creation of cyberperformances or teaching classes. We realized that Zoom, which is mainly aimed at business communication, has several limitations when it comes to artistic creation. One of them concerns the fact that we couldn't play musical instruments while singing because the system filtered the sound of the music by considering it background noise. At the same time, the system prevented any kind of overlapping voices that we would have liked to apply to the reading of the poems. As result, vocal strategies such as the canon or the choir, for example, had to be avoided. In addition to these sound constraints, the use of filters with the camera proved to be ineffective because what was visible on computer screen was different from what was projected. The differences in resolution of the projection equipment, as well as the nuances of light at the performance venue, were therefore uncontrolled variants.

This performance, as an experiment, thus underline one of the main conclusions we had already reached through research during the CypeT project, namely, the valuable inclusion of cyberperformance in curricula, but also the need to adapt online communication resources so that they can be effective in pragmatic communication and also enhance creative uses.

Acknowledgments. This work was supported by the Portuguese Foundation for Science and Technology (FCT) in the framework of the CyPet Project - EXPL/ART-PER/0788/2021- "Development of a new pedagogical model for teaching cyberperformance in higher education"- An initiative of the Center for Research in Arts and Communication (CIAC) - University of Algarve in partnership with the Open University of Portugal and the University of Maia. In this Congress we have support from CITEI - University of Maia. **fct** Fundação para a Ciência e a Tecnologia

References

1. Brönstrup, C.: Teatro virtual e nostalgia da presença. In: Fagundes, P., Dantas, M.F., Moraes, A. (eds.) Pesquisa em Artes Cênicas em Tempos Distópicos: rupturas, distanciamentos e proximidades [livro eletrônico] / [org.]. UFRGS, Porto Alegre (2020)
2. Flores, M.A., Machado, E.A., Alves, P., Vieira, D.A.: Ensinar em tempos de COVID-19: um estudo com professores dos ensinos básico e secundário em Portugal. Revista Portuguesa de Educação **34**(1), 5–27 (2021). https://doi.org/10.21814/rpe.21108
3. Flores, M.A., Gago, M.: Teacher education in times of COVID-19 pandemic in Portugal: national, institutional and pedagogical responses. J. Educ. Teach. **46**(4), 507–516 (2020)
4. Marques, D.: O impacto da pandemia de Covid-19 na digitalização do Ensino Superior. Dissertação de Mestrado em Gestão de serviços. Universidade do Porto, Porto (2021)
5. Gomes, I.M.S.: A digitalização do teatro em Portugal e a pandemia de COVID-19. Tese de mestrado em Comunicação, Cultura e Tecnologias da Informação. Instituto Universitário de Lisboa, Lisboa (2021)
6. Cunha, M.M., Maia, S.V.: Dar voz as artes: experiências artísticas em tempos de pandemia. In: Emoções, Artes e Intervenção- Perspetivas Multidiciplinares. Almedina, Coimbra (2021). http://hdl.handle.net/10316/97588
7. Tussi, G., Neves, E., Fásvero, A.: Aprendizagem criativa e formação docente no Ensino Superior. Revista Educar Mais, vol. 6, pp. 737–747, p.740 (2022). https://doi.org/10.15536/reducarmais.6.2022.2859

8. Resnick, M.: Jardim de infância para a vida toda: por uma aprendizagem criativa, mão na massa e relevante para todos. Trad.: Mariana Casetto Cruz, Lívia Rulli Sobral. Porto Alegre: Penso (2022)

9. Gomes, C.: Ciberformance: a performance em ambientes e mundos virtuais. Tese de Doutoramento em Ciências da Comunicação. Universidade Nova de Lisboa, Lisboa (2013)

10. Gomes, C.: Ciberformance: performance e netactivismo. In: Netactivismo. Org.: Isabel Babo, José Bragança de Miranda, Manuel José Damásio e Massimo Di Felice. Lisboa: Edições Universitárias Lusófonas. Colecção: Imagens, Sons, Máquinas e Pensamento, pp. 308–324 (2017)

11. Jamieson, H.V.: Adventures in Cyberformance - Experiments at the interface of theatre and the internet. Master's thesis in Drama, Creative Industries Faculty, Queensland Universty of Technology (2008)

12. Duarte, S.: Ciberformance, Second Life e Synthetic performances. Ouvirouver: Uberlândia, vol. 12, no. 2, pp. 448–460, ago/dez (2016)

13. Najima, F.M.: Ciberperformances e a Cibernética. Revista outras Fronteiras. Cuiabá-MT, vol. 7, no. 1, pp. 1–19 (2020). ISSN 2318-5503

14. Creely, E., Danah, H.: Creativity and digital technologies. In: Peters, M.A., Heraud, R. (eds.) Encyclopedia of Educational Innovation. Springer, Singapore (2019). https://doi.org/10.1007/978-981-13-2262-4_143-1

15. Santana, C., Moreira, J.: Cartografando experiências de aprendizagem em plataformas digitais: perspectivas emergentes no contexto das pedagogias das conexões. In: Espaços de aprendizagem em redes colaborativas na era da mobilidade. Org. Simone Lucena, Marilene Batista da Cruz Nascimento, Paulo Boa Sorte–Aracaju/SE: EDUNIT (2020)

16. Murray, J.H.: Hamlet on the Holodeck, The Future of Narrative in Cyberspace. MIT Press, Cambridge (2017)

17. Sapucaia, R., Vieira, C., Santos, I.: Poestics in cyberperformance: a case study on Jardins esquecidos. In: V Seminário Internacional de Pesquisa em Arte Visual (SIPACV (en) volver, pp. 461–474 (2023)

18. Bryman, A.: Social Research Methods. Oxford, New York (2012)

19. Scotti, V., Aicher, A.L.: Veiling and unveiling: an artistic exploration of self-other processes. Qual. Inq. **22**(3), 192–197 (2016)

20. Seregina, A.: (forthcoming) Performing Fantasy and Reality in Contemporary Culture. Routledge, London (2018)

21. Rolling, J.H.: A paradigm analysis of arts-based research and implications for education. Stud. Art Educ. **51**(2), 102–114 (2010). http://www.jstor.org/stable/40650456

22. Taylor, B.: Collage: The Making of Modern Art. Thames and Hudson Ltd, London (2004). 224pp

23. Wexel, J.: Teatro, audiovisual e *streaming*: uma análise sobre o fazer teatral em tempos de incerteza pandêmica na experiência pós-dramática da peça *Esperando Godette*. Rotura–Revista de Comunicação, Cultura e Artes (1), 39–46 (2021). https://doi.org/10.34623/sxc1-he90

24. Sapucaia, R.: Quintal dos sons: Um caminho para a imersão sonora. Tese de doutoramento em Média-arte digital. Universidade aberta em associação com a Universidade do Algarve (2022)

25. Bryant, J.: The Fluid Text: A Theory of Revision and Editing for Book and Screen. U. of Michigan Press, Ann Arbor (2002)

26. Bachelard, G.: A água e os sonhos: ensaio sobre a imaginação da matéria. Martin Fontes, São Paulo (1998)

27. Millôr, F.: Shakespeare Traduzido por Millor Fernandes. Porto Alegre: LP&M Editores. (1988)

28. Bauman, Z.: Amor Líquido: Sobre a fragilidade dos laços humanos. Jorge Zahar Editor, Rio de Janeiro (2004)
29. Lucena, S., Santos, V., Oliveira, A.: Espaços multirreferenciais de aprendizagem em redes colaborativas. In: Espaços de aprendizagem em redes colaborativas na era da mobilidade. Organizadores: Simone Lucena; Marilene Batista da Cruz Nascimento: Pauloa Boa Sorte. EDUNIT, Aracaju (2020)
30. Bidarra, J., Veiga, P.A.da., Sapucaia, R., Wexel, J., Tavares, M., Costa, S.: Desenvolvimento de um modelo pedagógico virtual paras as artes performativas digitais. RE@D - Revista de Educação a Distância e Elearning, vol. 6, Número 1 jan-jun (2023)

Bridging Computational Art and Climate Change: An Analysis of a Mobile Application for Raising Awareness About Climate Change Through Art

Felipe Mammoli Andrade[(✉)] and Artemis Moroni Sanchez Moroni

Cyberfisic Systems Division, Instituto de Ciência e Tecnologia (CTI) - Renato Archer, Rod. D. Pedro I, km 143, Campinas, SP, Brazil
{fmandrade,amoroni}@cti.gov.br

Abstract. This article focuses on the complex process of translating scientific data, specifically related to climate change, into comprehensible narratives aimed at engaging the non-expert public. Challenging the prevailing assumption that data possesses inherent communicative capacity, we analyze the complicated work involved in converting scientific data into meaningful visualizations and narratives. Furthermore, we investigate how artists have been contributing to transform the abstract aspects of climate change into tangible, emotionally resonant experiences, thereby cultivating novel forms of climate-aware public engagement. We then use this analysis to present and discuss the GaiaSenses Project, an innovative research project converging art, science, and computational technology. Through location-based audiovisual compositions driven by weather data, the project aims to establish a more intimate connection between individuals and the realities of climate change, fostering heightened environmental sensibility. By presenting and analyzing the GaiaSenses Project, this article reflects upon the transformative potential of computational creative systems in raising public awareness about climate change and stimulating climate action.

Keywords: computational art · climate change · climate art

1 Introduction

Contrary to the popular saying, data does not speak for itself. As numerous researches have repeatedly shown, making data speak involves quite a lot of work, specially for scientific data [1–3]. From designing beautiful visualizations and crafting short summaries for decision makers, to establishing trustful institutions for data validation and educating the audience to care about such data, making data speak also involves creating publics that are willing to listen to what data has to say. After all, does it really matter if data can speak for itself or not if there is no one there to listen to it? This simultaneous process of making data speak and making a public that cares about what it has to say is

A. L. Brooks (Ed.): ArtsIT 2023, LNICST 564, pp. 149–165, 2024.
https://doi.org/10.1007/978-3-031-55319-6_11

particularly clear in the history of the environmental sciences, specially those regarding climate change. As show by historian of science Paul Edwards in his study of the history of climatology and climate research institutions, the possibility to scale local data into a global reach phenomena that respects no territorial boundaries - i.e. climate change - was dependent on the capacity to build far reaching institutions that would make possible the material conditions of a global flow of weather and climate data, and knowledge [4].

Concretely, the possibility to collect weather data from around the planet, analyze it and publish reports like the IPCC Assessment Reports to foster policies like the Kyoto Protocol and Paris Agreement, is wholly dependent on the creation of institutions like the World Meteorological Organization (WMO), the Intergovernmental Panel for Climate Change (IPCC) and UNFCCC [4]. An analysis that echoes what Shapin and Shaffer (1987) have observed in their historical research on the debates between Thomas Hobbes and Robert Boyle over what means for an Air-pump to work and create vacuum, a debate that happened in XVII century Restoration England and came to be the birth of the experimental sciences: solutions to a problem of knowledge are solutions to the problem of social order [5].

While the enterprise of assuring scientific authority to the framing of climate change as a global phenomena has been largely successful from the perspective of climate sciences and its institutions, the same success has not been achieved in promoting significant political change to avoid the drastic futures presented since the 3rd Assessment Report [6, 7]. This political stagnation of climate policy is made clear by the failure of every country in the G20, the major emission offenders, in reaching any of the emission reduction goals since the Paris Agreement in 2015 [8]. Analyses like [6] have identified that for climate knowledge to be more policy-relevant there need to happen significant transformational changes in the structure and functioning of the main institution that produces scientific knowledge about climate change, the IPCC. And, due to the history of this institution, for these transformational changes to take place it needs significant pressure from both the inside, the scientists that are part of it, and from the outside, the society at large [6]. Non-expert people, an element that doesn't usually participate in the production of knowledge about climate change, but suffers the consequences of it nonetheless. This societal pressure, though, can only be exercised if the public can relate the issues raised by scientists with their common daily lives.

At least since the 90s, artists have been working with themes around climate change and global warming and trying to bridge the gap between the abstract and statistical knowledge about climate change and the experiential and daily lives of people [9]. These recent engagements of artists and the climate have been called by a varied set of names: climate change art, eco art, environmental art, climate art, atmospheric art, and many others. Despite the multiple designations, they all seem to share a commitment to raise awareness about the climate crisis and to foster an attempt to devise methods for inquiring the roles that the natural environment has in the human creative processes [9]. In other words, they strive to produce their publics [10].

Computers and computational systems have been crucial in these still early explorations. Weather by using computer systems to produce new media art exploring climate change issues, or using scientific data about the environment as material for artwork, artists dealing with climate change seems to have no option but to use computers as interaction and inspirational devices. In the myriad forms of engagement of artists with this digital objects, the abstract, statistical and global language of climate change, like "global mean temperature", "carbon budget" and "planetary tipping points", are transformed into a grammar of affect, closer to spatial and temporal scale of the human perceptual capacities. A transformation, as noted by the renowned artist Olafur Eliasson, that tries to make climate change a tangible phenomena [11].

In this text, we present and discuss the GaiaSenses project, an art-science project that also aims to make climate change more tangible for non-expert people. GaiaSenses is a mobile application that presents its users with visual animations and soundscapes based on the user's geolocation and the weather conditions of their surroundings. The animations and soundscapes are driven by weather and climatic data, like vegetation constitution, fire and rainfall, derived from NOAA multiband satellite images and made available by the Center for Meteorological and Agricultural Applied Research at University of Campinas - (CEPAGRI). By repurposing scientific environmental data into novel aesthetic compositions, the project aims to provide people with novel forms of engaging with important aspects of climate change that are not usually part of our weather experience, in the hope that it fosters more public engagement towards climate action. The GaiaSenses Project is being developed at the CTI Renato Archer by an interdisciplinary team of researchers. By presenting and discussing the GaiaSenses project this text aims to inquire into the capacities that computational creative systems can have in helping us make sense of climate change and reshape the ways we relate to the environment.

This article is composed of the following sections. First, we present a brief summary of how climate change has been made visible for both experts and non-experts by contextualizing the most famous data visualization, the hockey stick, in terms of computational creativity to show that science also depends on the aesthetic production of images for their comprehension and communication. Secondly, we present two important artworks that use different strategies to make the presumed distant climate change future closer to the public: Ice Watch by Olafur Eliasson and Oceans in Transformation by Territorial Collective. We describe them and present a brief analysis of some of their affective aspects. Then, we present the GaiaSenses project, how it works and some of its compositions in an attempt to try to locate where creativity is understood to reside in the system. Finally we present a brief discussion over some effects of dealing with climate change for computational creativity research.

2   Making Climate Change Visible: Strategies from the Sciences

Climate change is a paradoxical phenomena. While there is not a single place in the world one can go to see these changes in the global climate, there is hardly anywhere one can go to escape from them. This is due to the scale of the very object that is subject to these changes, the global climate. An entity that is way beyond the human scale of perception and can only be made visible by a vast network of environmental data collection devices distributed around the planet, like weather and marine stations, people with expertise to aggregate and process these data in meaningful ways, like climatologists and ecologists, and institutions that make possible both the flow of data and knowledge about the global climate, like the World Meteorological Organization (WMO) and the

Intergovernmental Panel on Climate Change (IPCC). Historian of science Paul Edwards called this arrangement the "knowledge infrastructure of climate change", as he defined it as the "robust networks of people, artifacts, and institutions that generate, share, and maintain specific knowledge about the human and natural worlds" [7, p. 340]. Due to the complicated nature of the phenomena, climate change is a subject in need of particular perception and cognition support to be minimally comprehensible [12]. And this support usually comes in the form of images and charts that help narrate the climate data [13].

We would be wrong to assume that only artists engage in the aesthetic reconstruction of climate data in more visually appealing formats. In fact, there is hardly a climate change image more famous than the "hockey stick" graph (Fig. 1). First published by climatologist Mann, Bradley and Hughes in 1998 [14], the graph depicts a statistical reconstruction of the planet surface temperature variation in the last 1000 years for the Northern Hemisphere, from year 1000 to the year 2000, based on climate proxies like ice cores, corals and tree rings. The figure that resulted from the complex statistical analysis was, as Mann describes in his book, "a wiggly curve documenting past temperature changes over the entire Northern Hemisphere (the hemisphere with the most data) and indicating a sharp rise in temperature over the past century" [15, p. xvii].

Even though Mann comments that he considers that the hockey stick graph was probably the least scientifically interesting thing in the article, it was that image that spawned massive media coverage [15, p.49]. The hockey-stick was featured in Vice President Al Gore's speeches in 1999, in president Bill Clinton State of The Union speech in the 2000, in the 2001 IPCC Assessment Report for Policy Makers and in 2006 Al Gore's An Inconvenient Truth movie. But the figure actually became globally famous because it was a main point of attack for climate deniers in the early 2000s, so much so, that in 2005 the chairman of the UK House Committee on Energy and Commerce and the chairman of the Subcommittee on Oversight and Investigations demanded an investigation of the full records of the climate research, personal finance and career of Mann and his co-authors. A request that was accepted by the National Academy of Science and after investigations, came to the conclusion that nothing was wrong with the research [15, 16]. Even with this conclusion, the same researchers were victims of the 2009 email hacking event in the University of East Anglia in the United Kingdom that leaked thousands of emails and documents exchanged between climate researchers that further spawned independent investigations over the conduct of climate scientists working on climate change [17].

Mann comments that he is not sure why that specific image was the one that stirred up so much controversies that caused him to suffer incessant harassment and threats from climate deniers and oil lobbyists, since paleoclimatic reconstructions that shown a rapid increase in temperature in recent times had been done before. But he suspects that it was the high resolution of time in the graph that they managed to achieve due to his and his co-authors innovative use of proxy climate data coupled with the statistical technique of principal component analysis (PCA). While previous researches present time resolutions of centuries or decades at best, Mann's research had managed to present the temperature variations year by year, a time scale very much in line with human perception of time [15]. Which made it possible to show that it is not just that the 20th century has been warmer than the last few centuries, but that the year 1990 has been the warmest year of

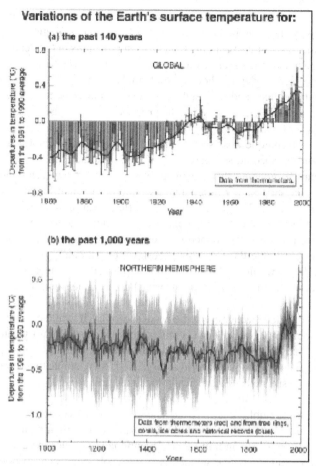

Fig. 1. Variations of the Earth's surface temperature over the last 140 years and the last millennium. (a) The Earth's surface temperature is shown year by year (red bars) and approximately decade by decade (black line, a filtered annual curve suppressing fluctuations below near decadal time-scales). There are uncertainties in the annual data (thin black whisker bars represent the 95% confidence range) due to data gaps, random instrumental errors and uncertainties, uncertainties in bias corrections in the ocean surface temperature data and also in adjustments for urbanization over the land. Over both the last 140 years − 0.8 and 100 years, the best estimate is that the global average surface temperature Year has increased by 0.6 ± 0.2 °C. (b) Additionally, the year by year (blue curve) and 50 year average (black curve) variations of the average surface temperature of the Northern Hemisphere for the past 1000 years have been reconstructed from "proxy" data calibrated against thermometer data (see list of the main proxy data in the diagram). The 95% confidence range in the annual data is represented by the gray region. These uncertainties increase in more distant times and are always much larger than in the instrumental record due to the use of relatively sparse proxy data. Nevertheless the rate and duration of warming of the 20th century has been much greater than in any of the previous nine centuries. Similarly, it is likely that the 1990s have been the warmest decade and 1998 the warmest year of the millennium. Source: IPCC AR3 Working Group I: The Scientific Basis p. 3, 2021.

the last thousand years. A picture that made every more clear the anthropogenic aspect of global warming by showing, in the human scale of time, the geological effect of human actions over time. A picture so enticing that, as Mann recounts, it had featured on a large number of magazines, articles, as cover art and as art in protest banners [15].

As the controversy over the hockey stick makes clear, it was the image and its aesthetic capacity to affect, and not just the climate data, that spawned the heated debates. Although Mann describes in some detail the process of producing the graph, how it was slowly composed by incorporating climate data that were already collected elsewhere and how the PCA technique was used to make all the data interoperable, the author does not comment on the specific software tools used to produce the image. Nonetheless, it is not too hard to imagine that the image was created using techniques similar to the ones used today in programming languages that offer scientific visualization tools, like Mathlab, R, Python. By appropriating [18] categorization of computational art for analyzing this particular scientific image, a categorization that describes computational art as "a spectrum of activities defined by the margin of control of human artist on the final output, or, to reverse the viewpoint, the level of autonomy of the computational system" [18, p157], it would not be too far off to suppose that the hockey stick belongs to the leftmost side of the spectrum, where an artist have complete control over the process and uses code as medium.

In the terms of Boden's cognitivist theory of creativity, it would be hard to determine if the hockey stick would better fit the combinatory or the exploratory category, but in a retrospective analysis it had a transformational effect in both the conceptual space of climate science and in the societal world at large [19]. An example that reframes categorization theories as situated and historically localized practices [20]. As [17] comments, these controversies had a significant effect on the public trust in science and surveys showed a decrease in public trust in what scientists have to say about the environment both in the US and Europe [17, p548]. An important transformational effect however, was in the worldview of Mann himself. He mentioned that after the events, it became clear to him that climate change science is inextricably connected to politics [15]. A point that many artists and scientists try to make clear to non-expert publics.

The hockey stick story shows the significant societal effect that aesthetic experimentation with climate data can have, even when inside the aesthetic rules of scientific representations. In the next section, we present two artworks that deal directly with climate change and discuss some aesthetic and technical strategies that they use to try to bring the planetary scale of climate change down to the human scale of perception.

3 Making Climate Change Tangible: Strategies from the Arts

3.1 Making the Distant Close - Olafur Eliasson's Ice Watch (2015)

The monitoring of ice sheets is a main index to assess climate change. The extent of ice sheets in the poles, especially in Greenland and Antarctica, fluctuate naturally due to seasonal changes and weather variations, like temperature, precipitation and humidity. The expected behavior is that melted ice would drain into the sea, later evaporate and, due to the water cycle, would eventually fall as snow atop the ice sheep, keeping its amount in a dynamic balance. But with climate change, the planet warms due to the continued emission of CO2 worldwide, making the warm seasons longer, thus interrupting the time the melted ice needs to go back to its ice sheet. This accumulation of lost ice is an important metric to assess sea level rise worldwide, one of the main risk factors of global climate change. Since 1992 Greenland and Antarctica have each lost more than 100 billion metric tons of ice per year on average [21].

This monitoring relies on complex aggregation of historical satellite imaging and data from marine buoyancy stations and is usually visualized in line charts depicting the decline of ice sheet mass over time. Despite being a central risk element for populations worldwide, it is a phenomenon that is happening in very remote places, like Greenland and Antarctica, in the glacial landscapes that are mostly unknown for non-expert people and at a scale particularly hard to comprehend. It is as an attempt to raise awareness about these often remote and abstract phenomena that characterize global climate change that the artist and United Nations (UN) Climate Ambassador Olafur Eliasson presented the installation Ice Watch in 2015.

Ice Watch is an interactive installation developed by the Danish-Icelandic artist Olafur Eliasson in collaboration with geologist Minik Rosing and presented in Paris for the occasion of the 21 COP Climate Summit in 2015 and subsequently reprised in London and Copenhagen. In 2015, the installation consisted of twelve immense blocks of glacial ice arranged in a clock formation at the Place du Panthéon in Paris, and left to melt away during COP21.

From December 9th to December 14th 2021, while the talks that would later become the Paris agreement were happening inside COP21 and the ice melted, the passing by people were encouraged to interact with the ice blocks the way they see fit, in an effort to inspire public action against climate change. A large number of people were drawn by the intriguing blue hue of the glacier ice and the chance to see it and contemplate it [22] (Fig. 2).

A number of photos show people interacting with it in a myriad of ways, touching it, hugging it, sitting at it, drinking the melted ice water, taking pictures, dancing near it and even moving it as it becomes smaller and lighter. As reported by the writer Rebecca Solnit and recovered by [9], "it's a beautiful, disturbing, dying monument to where we are right now... People are coming by fascinated, most needing to touch the ice" [9, p. 144]. A response that seems to be aligned with the artist's original expectation in the press release: "Put your hand on the ice, listen to it, smell it, look at it – and witness the ecological changes our world is undergoing. [...] I hope that Ice Watch arouses feelings of proximity, presence, and relevance, of narratives that you can identify with and that make us all engage" [24].

As [9] notes, the way the ice fragments established a relationship with the broad theme of climate change is not just due to its origin in the receding glacier of Greenland, but also due to its trajectory to the exhibition. Weighting a total of 80 tons, the twelve blocks were harvested from free-floating icebergs in a fjord near Nuuk in Greenland by divers and dockworkers from the Royal Arctic Line. They were then placed in six 40 foot refrigerated containers in Nuuk, Greenland, and transported to Aalberg in Norway by a container ship. In Norway, the containers were moved to Paris by truck and arranged at the plaza using forklifts and a crane [9]. A journey that amounted to a total of 30 tons of CO2e in carbon footprint [Olafur emissions report]. The very physical transportation of

Fig. 2. Ice Watch Installation in Paris. Source: Ice Watch Paris Press kit.

the key actors, the ice blocks, to a place of decision about the global climate as the COP21 performs the whole issue of climate change. As they travel, they get enmeshed in the global logistical network that connects every part of the planet, and that is significantly responsible for global CO2 emissions [9]. At the end of the journey, the ice blocks are a kind of 10 ton miniatures of global climate change.

When the public interacts with the melting ice, the geological scale of climate change and the human scale of sensing collapse onto each other. By breathing next to the ice blocks, one breaths the ancient air trapped between the snow layers that form the glacier ice, by touching it, one is made part of the milenia long process that forms glaciers by the slow deposition of snow that is compressed by its own weight, and by witnessing them melt, one can see, like in a natural time lapse video, the future that awaits humanity in the next century [9]. In Ice Watch, the control that the artist has over the final artwork is distributed to the weather forces of both the surrounding environment and surrounding publics, like the heat from the atmospheric temperature and the heat emanating from people's body.

The installation works not only because the artist provides unmediated access to a rare natural phenomena, the melting of glaciers, but because the artist was capable of arranging a complex set of technical mediations that allows for the ice blocks to be moved from Greenland to Paris before it fully melts. The efficacy of this artwork lies, of course, in this capacity to transport, alongside the ice block, the environmental relations that would only happen in a remote place, like Greenland, to a crowded place like Paris. And at a crucial time for climate politics, the COP21.

The Ice Watch does not seem to be closely related to computational art at first, but it can be thought of as an ingenious way of visualizing fundamental data about the changing climate, the melting of glaciers, but directed at non-expert publics. It functions as an analogy to climate scientists watching the planet change inside their computer models by analyzing data tables and charts. But instead of affecting its publics by a "visual analytics" [25] like the hockey stick graph, it affects its publics by making concrete the abstract phenomena of melting glaciers and bringing close to home the distant research sites in the coldest parts of the northern Atlantic Ocean. These aesthetic strategies are central to the objective that Olafur Eliasson sets for his own work, as he comments: "As an artist I hope my works touch people, which in turn can make something that may have previously seemed quite abstract into reality [...] and Ice Watch makes the climate challenges we are facing tangible" [22]. The expectation, as the author comments, is that by allowing this intimate connection to be forged, the experience of connection will inspire public action.

3.2 Making the Unimaginable Imaginable - Territorial Agency's Oceans in Transformation (2020)

With Ice Watch, Olafur Eliassion hoped to forge connections between the individuals and the planet by providing an intimate sensory experience to an index of climate change that is hard to imagine from a distance. In Oceans in Transformations (2020), the architecture collective Territorial Agency tackles the difficulty of imagining our climate future using a different set of techniques. Instead of bringing a miniature of climate change, Oceans in Transformation brings together numerous datasets of human effects on the planet into a single map, providing an impactful picture of modern history and the future to come.

Oceans in Transformation is an art-research project that traces the human impact in the oceans by collecting and collating data from 79 different datasets in one large map that tries to show a picture of what is known about the oceanic system and what it could look like in the future [26]. The project was developed by the architecture collective Territorial Agency, founded by John Palomino and Ann-Sofi Rönnskog and it was commissioned by the Thyssen-Bornemisza Art Contemporary (TBA21) - Academy. The three year project culminated in an exhibition in 2020 at the Ocean Space at the old Church of San Lorenzo in Venice, Italy, and was the recipient of the S + T + ARTS'21 prize promoted by the European Commission for project where appropriation by the arts has a strong potential to influence or alter the use, deployment, or perception of technology [27].

In Venice, the core of the exhibition was an installation composed of 37 large LCD panels distributed in 7 groups (Trajectories), where each group depicted a large animated map of a transect of the planet and the numerous human activities that happened in this zone [28].

The grouped LCD panels depict maps that the artists called the "sensible zone" [28], a small vertical section of the planet that spans from 200 m below the ocean level to 200 m above the ocean level. A section that concentrates the most biogeochemical activities - life - of the planet's biosphere, the section that historically contains the highest level of human activity and environmental impact, and coincidentally, the section that will be most affected by the 3 m–6 m rise in the sea level projected for the next century.

In one of the groups, the North Sea to Red Sea Journey (Fig. 3), a transect cuts from the North Atlantic, across the continental shelf of the North Sea (passing through

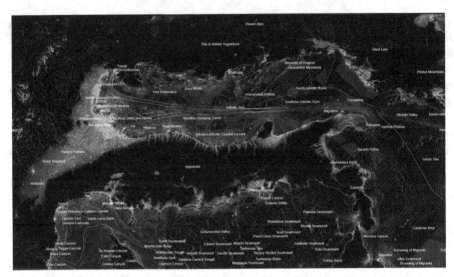

Fig. 3. An example of the visual composition present in the exhibition Oceans in Transformation. In this case, the sensible zone of Italy. Source: [26]

the UK, France and Italy), it goes through Venice and its lagoon, across the Adriatic, it connects to the Suez Canal and finally ends at the Red Sea. The transec shows an enormous map, where only the sensible zone (200 below and 200 above the sea level) is colored and the rest is dark. In this area, 79 different datasets with data ranging from 1900 to 2020 are superimposed considering the geolocation of their data, like a long exposure photograph taken from a satellite over 120 years [26]. In the map, the green territories are land from sea level to 200 m above it, the white lines are the trajectories of transport ships, the tiny white dots are fishing ships, red squares and circles are oil and gas exploration stations. On the land, the different intensities of light represent the urban expansion of cities, and in their interface with the sea, the white to light blue to dark blue color gradient represent the projected sea level rise for climate scenarios of + 1,5 °C, +2 °C and +3 °C above pre industrial levels. And the sea rise distribution is a result from sea rise projections combined with a digital elevation model DEM derived from the Shuttle Radar Topography Mission SRTM [28].

The Ocean in Transformation project uses only open access data and a list of the 79 datasets used can be found in the project website [26], along with the specific portion of the data that was used. As commented by John Palomino in an interview about Oceans in Transformation, although all data was open access, downloading them and making them interoperable was incredibly laborious and time consuming, since the data sets spanned through numerous different scientific fields that do not share data standards between them and rarely had the same space or time resolution. As shown by the artists, knowledge about the ocean is still incredibly fragmentary, with large portions of the oceans still unstudied. The little data available, observes John Palomino in an interview, is scattered in many different websites that are usually known only by experts in the field and systems with complicated processes of data requests [29].

The aesthetic language born out of Oceans in Transformations has the purpose of making visible the enormous human impact in a space usually associated with emptiness, the ocean, by making the most of the land not visible. As Fig. 3 shows, the ocean is not empty, on the contrary, it is incredibly busy with human and non-human activities, and incredibly at risk. And by bringing together many disparate datasets about the oceans, it makes it possible for the public to grasp, visualize and imagine the future of the crucial, but often neglected oceanic system. By seeing the world from the point view of the oceans, from the point view of the sensible zones, it is possible to comprehend that we, humans, are actually part of a collective that shares similarities by means other than land.

Differently from the Ice Watch, the control the artist has in the final product is extreme, but this extreme control is used to highlight the gaps in our knowledge about the ocean by showing that despite this control, it is the data availability that defines the possible aesthetic composition in the panel. Gaps in data become gaps in the knowledge that become gaps in the possibility to fill the visual composition.

By analyzing the two artworks above, Ice Watch (2015) and Oceans in Transformation (2020), it is possible to extract two principles that could contribute to the effectiveness of creating climate change artwork with computational means. First, based on Ice Watch (2015), the notion that no matter how precise the scientific data about changes in the environment are, they are usually understood as too abstract by the non-expert public, since the global climate is a non-tangible object. Thus, it is crucial to foster other ways to visualize climate data that are more embodied and allow for sensible connections, as the artists observe in their essay accompanying the artwork: "there is a huge gap between what we know and what we feel. How can we translate knowledge into action, and really change our behavior? Of course, it is necessary to present the facts and data supporting climate change science, but this is not where action begins. Only by embodying knowledge can we gain a sense of responsibility and commitment" [23, p4].

In Oceans in Transformation (2020), the artists make clear that the usual cartographic imaginary that guides our view of the world is not enough to comprehend or even imagine the extent of the human impact on the planet, especially in the oceans. And that the fragmented aspect of knowledge about these impacts contributes to this alienation. To actually grasp the extent of the magnitude of climate change in the world, it is crucial to develop new aesthetic propositions that offer alternative perspectives on the environments we inhabit. And as they show, these alternatives do not need to abandon scientific knowledge and procedures, on the contrary, the extensive use of tools like Geographic Information Systems (GIS), Satellite Images and ocean datasets, can actually open possibilities to experiment with environmental data in novel aesthetic ways.

4   The GaiaSenses Project

The previous sections have shown related works of climate art that try to raise awareness about the changing climate by providing an aesthetic experience for the public in the form of an event. In the GaiaSenses project, the strategy is to provide its public with periodic aesthetic experiences so they can perceive the climate and the weather in different ways, making aspects of climate change more present in their daily life.

The GaiaSenses Project proposes the development of an application for mobile devices that periodically delivers audiovisual compositions to its users, based on the weather conditions of their surroundings. Based on the users geolocation data, the application will periodically generate an animation and an accompanying sound using data from the weather satellite GOES-16, made available by the Center for Meteorological and Agricultural Applied Research at University of Campinas (CEPAGRI) and data platforms like Google Earth Engine and OpenWeather. These compositions will be short videos (20–30 s) or interactive images bringing attention to specific weather and environmental events happening near the user's location, like rain, fires, or vegetation composition, and users will be able to share these compositions online in social media platforms. By providing a periodic composition, the project aims at making climate change a more present element in the daily life of people, a topic that is not only remembered in moments of environmental disaster [30].

The audiovisual compositions will be made by using these localized data as inputs to generative algorithms and other creative computing techniques. Currently, the visual components are produced by specific animation algorithms implemented using the javascript library p5.js, a library for creative coding similar to Processing. The project currently has 12 visual animations implemented that are powered by real time weather data such as temperature, quantity of rain, lightning occurrences, wind speed, wind direction, cloudiness and fire occurrences. Each animation receives a combination of these weather data that modifies its behavior. For example, Fig. 4, the animation Storm Eye: it receives temperature data, wind speed and wind direction. And they affect the animation as follows: 1) the temperature controls que color pallet of lines, from blue (cold) to red (warm); 2) the wind speed controls the velocity in which each curve circles around the screen; and 3) the wind direction controls how disperse the curves are from the center of the composition.

The audio compositions work in a similar way. The audio components are implemented in Pure Data, a visual programming language for creating interactive and real time music based on audio synthesis and processing, much like a modular synthesizer. The music compositions also respond to specific weather data that alters the composition behavior. In the Lluvia composition, for example, the amount of rain (mm/h) controls the pitch, tempo and audio channel, in a way that the heavies the rain one can hear a fast and spatial sound event, much like in the rain, where one can hear the droplets hitting the ground from every direction. Other examples of compositions can be seen on Fig. 5.

To produce the data to drive these animations, the project has a data processing service in a backend implemented in Python, using the Flask web framework, to leverage the rich data processing ecosystem available for Python. The data processing service works by downloading and transforming data from a desired location, it downloads Rainfall data from the OpenWeatherAPI, Fire data from the CSV files openly available in INPE website, and lightning data from the GOES-16 satellite also openly available in a AWS service. After downloading, the backend processes the raw data into metrics to determine what kind of animation will be produced for a specific location based on the seriousness of weather events happening in the requested region. This decision is based on extracting certain parameters from the raw data like the quantity of rainfall, fire temperature and lighting occurrence, and comparing them to thresholds [31, 32]. Thus, the user will

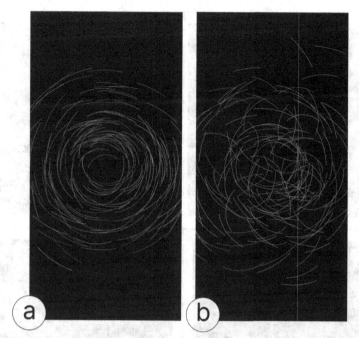

Fig. 4. Storm Eye composition in two different weather conditions. a) temperature: 24 °C, wind speed: 4 m/s, wind direction 20°; b) temperature: 32 °C, wind speed: 14.2 m/s, wind direction 271°. Source:

Fig. 5. Screenshots of GaiaSenses App showing some compositions. a) Brush, driven by humidity data; b) Lightning Tree, driven by lightning occurrences data; c) Zigzag, driven by rain and lightning occurrence data; d) Lluvia, driven by rain data. Source:

receive a composition that corresponds to the most serious weather event happening in their requested location, fostering a sense of local relation to the climate and weather systems.

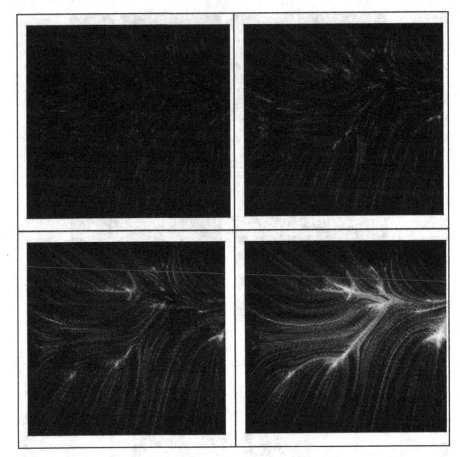

Fig. 6. Frame sequence of an animation using lighting data. Source: [31]

Figure 6 shows a sequence of frames of an animation driven by lightning data. In this animation, the data processing service downloads lightning data from the GOES-16 AWS repository, that make files available in netCDF format, the netCDF file is transformed into a CSV table containing latitude, longitude and the date and time of lightning occurrence, then, the CSV file is filtered according to the requested location and time period. Finally the data is sent to the frontend application that renders the animations in the user's mobile device.

In Fig. 6, the animation prints in the user's screen the location of every registered lightning event in a way that the more alarming the lightning event in a place, the more attracted they will be to each other, forming the bright areas seen in the image.

In [18] computational art autonomy spectrum, the GaiaSense is probably in the leftmost section, where the artists retain control over the output and use code as a medium. But when accounting for the decision making process and specific visual content of each animation and soundscape, the creative decisions are not made by the programmers alone, but also by the specific natural conditions, collected by weather satellites like the GOES-16 and the animation algorithms, thus distributing the creative agency of the artist/programmer over the final output to non-human elements like the environmental forces and the algorithms driving the animations.

Similar to the Ice Watch, the GaiaSenses provides a novel way to make climate data perceivable in the scale of the human senses, mostly by vision and hearing. And similar to the Oceans in Transformation, by joining previously dispersed data sources, it creates an aesthetic grammar to help people to engage intellectually and imaginatively to the often dry, but important climate data. A development that displaces the autonomous scale of computational art proposed by [18] at least a little to the right, towards a more dynamic integration between human and machine creativity.

5   Discussion and Future Work

This article presented the GaiaSenses project and discussed it in context to other instances of climate art created using computational means. By presenting other forms of aesthetically experiencing climate change, from more to less scientifically inspired, we intended to contribute to the still emerging discussions of how to use digital and computational art to raise awareness about climate change and inspire political action. In the discussed artworks, Ice Watch and Oceans in Transformation, we identified certain strategies that could be fruitful as reflexivity elements in the development of computational art dealing with climate change, and particularly in the GaiaSenses project. Particularly the principal of making abstract scientific scales more concrete and bringing together disperse scientific knowledge in creative aesthetic forms. As a result, we suggest that: as the advent of computing has prompted new ways of understanding and inquiring about human creativity [19], the acknowledgement that climate art makes about the active role that the environment has in making art, can make important advancement in accounting for more elements in the human creativity, beyond the psychological and social aspects. In an analogous way, the agency that the environment takes from us in this new climate condition, showing that nature is not just the passive background of human history, is similar to the dislocation of cognition that the computer has caused, claiming that cognition is not something that happens in isolation, only inside the the mind. It is, of course, extremely early to propose such ideas, but they seem relevant to at least touch on, due to not only the climate road ahead, but to our climate present.

As noted by [33], the computational creativity field has yet to engage with the matter of climate change, since there are very few people in the field tackling such relevant themes [30, 34]. According to [33], this engagement is a great opportunity since the field has historically focused on aspects that are now crucial to deal with the climate crisis, like supporting and enhancing decision-making processes, automatically generating creative scenarios and promoting interdisciplinary partnerships with artists, scientists and decision makers. How the field will be able to contribute, as the authors notes, is now a matter of discussion, research of trial and error. A methodology that the computational creativity field seems particularly apt to execute and a topic that it will be less and less likely to be avoided.

Acknowledgements. We would like to express our gratitude to the researchers from CEPAGRI, Dr. Priscila Coltri, Dr. Renata Gonçalves, and Dr. Jurandir Zullo Jr., as well as M.Sc. Bruno

Bainy and Engineer Bruno Veloso, for their support of this proposal and the valuable seminars they conducted on meteorological data. This proposal also includes the participation of Prof. Dr. Amílcar Cardoso and Dr. Pedro Martins from the University of Coimbra, Prof. Dr. Jônatas Manzolli (IA-NICS) from Universidade de Campinas (UNICAMP), and Dr. Josué Ramos from DISCF/CTI Renato Archer. Students from various departments have been contributing to the development of GaiaSenses, we thank Conselho Nacional de Desenvolvimento Científico e Tecnológico (CNPq) for their scholarships, including Cássio Dezotti, Gabriel Kuae, Tauane Cardoso, and Ru Yi Shen from the Control and Automation Engineering Faculty; Elton do Nascimento, Thiago Lacerda, and Isabella Rigue from the Computer Engineering Institute; Gabriel Dincao from the Music Department; Álvaro Costa, Sara Freitas, and Pedro Trama from the Information Systems program, all of them from UNICAMP. We also thank CNPq for Felipe Mammoli Andrade PCI grant number 304295/2023-0. The GaiaSenses proposal is supported by CTI Renato Archer, an institute from the Ministry of Science, Technology and Innovation, and is currently being submitted to funding agencies.

References

1. Leonelli, S.: On the locality of data and claims about phenomena. Philosophy Sci. **76**(5), 737–749 (2009)
2. Boyd, D., Crawford, K.: Critical questions for big data: provocations for a cultural, technological, and scholarly phenomenon. Inf. Commun. Soc. **15**(5), 662–679 (2012)
3. Leonelli, S., Beaulieu, A.: Data and Society: A Critical Introduction. SAGE Publications Ltd., London (2021)
4. Edwards, P.N.: A Vast Machine: Computer Models, Climate Data, and the Politics of Global Warming. MIT Press (2010)
5. Shapin, S., Schaffer, S.: Leviathan and the Air-Pump: Hobbes, Boyle, and the Experimental Life. Princeton University Press (1985)
6. Asayama, S., De Pryck, K., Beck, S., et al.: Three institutional pathways to envision the future of the IPCC. Nat. Clim. Chang. **13**, 877–880 (2023)
7. Edwards, P.N.: Downscaling: from global to local in the climate knowledge infrastructure. In: Harvey, P., Jensen, C.B., Morita, A. (eds.) Infrastructures and Social Complexity: A Companion. Taylor & Francis (2016)
8. Holz, C.: Are G20 countries doing their fair share of global climate mitigation? Comparing ambition and fair shares assessments of G20 countries' nationally determined contributions (NDCs), Oxfam International, Oxford, September 2023
9. Randerson, J.: Weather as Medium: Toward a Meteorological Art. MIT Press (2018)
10. Noortje, M.: Material Participation: Technology, the Environment and Everyday Publics. Palgrave, Basingstoke, UK (2012)
11. Eliasson, O.: Ice Watch Homepage. https://icewatchparis.com
12. Windhager, F., Schreder, G., Mayr, E.: On inconvenient images: exploring the design space of engaging climate change visualizations for public audiences. In: Bujack, R., Feige, K., Rink, K., Zeckzer, D. (eds.) Workshop on Visualisation in Environmental Sciences (EnvirVis), pp. 1–8 (2019)
13. Dourish, P., Gómez Cruz, E.: Datafication and data fiction: narrating data and narrating with data. Big Data Soc. **5**(2) (2018)
14. Mann, M., Bradley, R.S., Hughes, M.K.: Global-scale temperature patterns and climate forcing over the past six centuries. Nature **392**, 779–787 (1998)
15. Mann, M.E.: The Hockey Stick and the Climate Wars: Dispatches from the Front Lines. Columbia University Press (2012)

16. Besel, R.D.: Opening the "black box" of climate change science: actor-network theory and rhetorical practice in scientific controversies. South Commun. J. **76**(2), 120–136 (2011)
17. Lahsen, M.: Climategate: the role of the social sciences. Clim. Change **119**(3–4), 547–558 (2013)
18. Daniele, A., Song, Y.Z.: AI+art = human. In: Proceedings of the 2019 AAAI/ACM Conference on AI, Ethics, and Society, pp. 155–161 (2019)
19. Boden, M.A.: Computer models of creativity. AI Mag. **30**(3), 23 (2009)
20. Sorting Things Out: Classification and Its Consequences. Geoffrey C. Bowker and Susan Leigh Star. MIT Press, Cambridge, MA (1999)
21. Rignot, E., Mouginot, J., Scheuchl, B., van den Broeke, M., van Wessem, M.J., Morlighem, M.: Four decades of antarctic ice sheet mass balance from 1979–2017. PNAS **116**(4), 1095–1103 (2019)
22. Eliasson, O.: Ice Watch Paris Homepage. https://icewatchparis.com. Accessed 21 Sept 2023
23. Rosing, M., Eliasson, O.: Ice, Art, and Being Human. Studio Olafur Eliasson (2015)
24. Eliasson, O.: Ice Watch London Homepage. https://icewatchparis.com. Accessed 21 Sept 2023
25. Coopmans, C.: Visual analytics as artful revelation. In: Coopmans, C., Vertesi, J., Lynch, M., Woolgar, S. (eds.) Representation in Scientific Practice Revisited. The MIT Press (2014)
26. Territorial Agency. North Sea to the Red Sea Trajectory homepage. https://www.territoriala gency.com/oceansintransformation-3/trajectory-north-sea-to-red-sea. Accessed 24 Oct 2023
27. Grand prize of the European Commission honoring Innovation in Technology, Industry and Society stimulated by the Arts homepage (2020). https://starts-prize.aec.at/en/oceans-in-tra nsformation/. Accessed 24 Oct 2023
28. Zyman, D. (ed.): Oceans Rising: A Companion to Territorial Agency: Oceans in Transformation. National Geographic Books (2022)
29. Territorial Agency Lecture at the Studium Generale Rietveld Academie. https://www.you tube.com/watch?v=R66h_QpeNPg
30. Moroni, A.: Data Art, Criatividade Computacional e Mudanças Climáticas. In: Rocha, C., Venturelli, S., Cruz, D. (eds.) Anais do IX Simpósio Internacional de Inovação em Mídias Interativas. Anais do 21o. Encontro Internacional de Arte e Tecnologia. 9th. Balance-Unbalance. ENTROPÍAS 2022. Universitad de Chile; Media Lab/BR, Santiago, Chile (2022). ISSN 2238-0272
31. Rigue, I., Moroni, A.: GaiaSenses: Acesso à base de dados de satélite e tratamento de seus produtos. Anais da XXIV Jornada de Iniciação Científica do CTI Renato Archer - JICC 2022. CTI Renato Archer, Campinas (2022)
32. Costa, A., Moroni, A.: GaiaSenses: o Frontend e Backend de uma aplicação móvel para a geração automática de composições audiovisuais baseadas em dados climáticos. Anais da XXIV Jornada de Iniciação Científica do CTI Renato Archer - JICC 2022. CTI Renato Archer, Campinas (2023)
33. Pease, A., Pease, A.: Computational creativity and the climate crisis. In: ICCC 2023 14th International Conference on Computational Creativity (2023)
34. Chang, J., Ackerman, M.: A climate change educational creator. In: ICCC, pp. 77–80 (2022)

Guitar Improvisation Preparation and Practice: A Digital-Assisted Approach Integrating Set Theory and Mechanical Gesture Exploration

Leon Salcedo[1,2]([✉])([iD]), Maria F. Zuniga-Zabala[3], and John A. Guerra-Gomez[4]([iD])

[1] Universidade de Aveiro - INET MD, Aveiro, Portugal
lfsalcedo@ucundinamarca.edu.co
[2] Universidad de Cundinamarca, Fusagasugá, Colombia
[3] BTactile, San Francisco, USA
mafe@btactile.com
[4] Northeastern University, Boston, USA
jguerra@northeastern.edu
https://btactile.com

Abstract. In this paper, we introduce a framework for improvisation in classical and contemporary guitar preparation. Our proposal follows a digitally assisted approach to support guitarists in practicing improvisation and creating coherent musical discourses. By simplifying the complex tasks involved in evolving a mechanical gesture and controlling harmonic development, the framework aims to enhance improvisation in the context of contemporary classical music. The paper addresses the challenges faced by classical guitarists in improvisation and emphasizes the significance of integrating mechanical memory and audition. We argue that guitarists can expand their harmonic palette and enhance their improvisational skills by following our digitally assisted approach that integrates critically Set theory principles and the exploration of mechanical gestures. The paper discusses the conceptual foundations of the framework, including the integration of Set theory and mechanical gesture exploration. It also outlines the contributions of the authors, which include the framework itself, computer modeling for the problem, and the implementation of an open-source tool for interacting with the framework. Our main goal in this work is to provide a valuable resource for guitarists interested in improvisation and to broaden the possibilities for creative expression in the realm of classical contemporary guitar performance. The interface can be tested on https://infovis.dev/viz/guitarImproviser.

Keywords: Classical contemporary guitar improvisation · Digital-Assisted Approach for music improvisation · Contemporary Classical music improvisation preparation · Guitar mechanical gesture exploration · Music improvisation using Set Theory

A. L. Brooks (Ed.): ArtsIT 2023, LNICST 564, pp. 166–185, 2024.
https://doi.org/10.1007/978-3-031-55319-6_12

1 Introduction

The scope of the work we present is mainly aimed at guitarists who carry out their artistic practice, in different styles of contemporary Western classical music stressing post-tonal music, however, we consider that the approach we present can be consulted, experimented with, and useful for guitarists and composers of other musical fields.

As noted by authors such as Després et al. (2017) [2] there is a growing interest in improvisation within the field of Western classical music, exemplified by the body of literature evidencing the positive impacts on music learning and also in the increasing presence of improvisation in concert halls and competitions.

However, in the learning and practice of performance in Western classical music consolidated mainly since the middle of the nineteenth century, the great majority of performers use their technical skills to memorize repertoires and their interpretative visions, mechanizing every detail and trying to be highly consistent in their performances. This is because the study and practice of Western classical music have been strongly oriented toward some principles such as the approximation from literacy, stylistic correctness, technical development focused on high precision and the search for virtuosity and the study of referents and authority figures such as fidelity to the score, canonical versions of the core repertoire, and adherence to certain schools of interpretation.

As a consequence, and if compared to other styles such as rock or jazz, classical music performers interested in improvisation face problems and insecurities to get started in the world of improvisation, as they feel they must develop a very broad and deep theoretical understanding of musical styles. This becomes much deeper in the classical contemporary era, that is to say, most of the twenty and twenty-first centuries, due to the great variety of individual styles developed by different composers. In addition, the performers' very practice and training focused on reproduction distances them from the necessary development of skills for improvisation. According to Hogan (2022) [6] recent research carried out by David Dolan and Henrik Jensen focused on scanning the brain activity of performers and demonstrated how reproductive memory and improvisation have two separate and very different brain functions. In opposition to this, the approach to improvisation in other styles such as rock and jazz is more intuitive by starting from a more direct link between mechanical memory and audition, as pointed out by authors such as Caporaletti (2018) [1], who call this approach the audio-tactile matrix.

In the field of the contemporary classical guitar repertoire, we find a large number of languages and styles that must be analyzed and understood in order to achieve similar musical results through improvisation and composition, which makes it challenging to proceed directly through an audio-tactile perspective.

Despite this, and bearing in mind that an audio-tactile approach to musical creation and especially to improvisation has proven to be so successful in styles and traditions such as jazz and rock to fluidly create idiomatic musical pieces and performances - in our case for the guitar -. We have wondered whether is it possible to propose a similar framework for creation, improvisation, and performance on contemporary classical guitar. And, if possible, how to achieve this, taking into account the number of languages and complexities that exist in the field of contemporary classical music and especially in post-tonal music?

One of the most developed analytical theories for the understanding of contemporary classical music is the Set Class Theory, proposed by Allen Forte in 1973 [3] and deepened by other theorists such as Joseph N. Straus in 1991 [10]. This theory allows synthesizing and rationalizing particular harmonic languages in the field of post-tonal music by finding and classifying groups of sounds that can be related by their intervocalic content, in other words, by the harmonic relationships of pitches that exist between the notes that are grouped in a given set.

Nowadays, there are digital tools that can be used to perform the necessary calculations for musical analysis through Set Class Theory; however, from an analytical point of view, these tools help to visualize the calculations to understand the interval content of a given set, but they do not synthesize and order results in a way that allows establishing and easily visualizing relationships of unity, transition, and contrast between sets of notes that allow developing a musical discourse. Much less, they do not offer the possibility of approaching the guitar from an intuitive perspective where audition and mechanical memory are linked in an audio-tactile way, that is to say, the way musicians (in our case guitarists) display the sets of notes in their instrument (guitar) and how they link this with their sense of audition in the course of a musical creation.

In general, there's a shared interest in the pursuit of new ways to understand improvisation and performance in classical contemporary music; and it is for that reason that we present in this work a framework involving conceptual and digital tools that enables one to understand how a set of notes on the guitar can be developed harmonically based on Set Class Theory through relations of unity, transition, and contrast, as well as how one can modulate the degrees of physical coincidence or similarity of the guitarist's mechanical gesture, that is, the displayed fingerings of these sets on the guitar. This enables a method of creating music where guitarists can put more of an emphasis on the intuitive experience of connecting mechanical memory and audition in order to experiment and build skills that will enable them to dive more deeply into the practice of improvisation.

In the following sections, we will present our conceptual approach and examine relevant literature pertaining to the different aspects of our proposal. Subsequently, we will delve into the theoretical foundations of our framework, encompassing the domains of harmonic development within the context of Set Theory and the evolution of the mechanical gesture. We will then outline the integration of these two components through computational modeling and its implementation within an open-source digital application. Additionally, we will provide an example illustrating the interaction between the framework and the digital tool, showcasing the generation of a musical piece through improvisation. Finally, we will discuss the conclusions drawn from this experience and outline potential future directions for this proposal. Drawing from the information presented in this paper, we can summarize our contributions as follows:

- Framework for the composition and improvisation of music for the guitar, based on an exploration of the link between mechanical (physical)

performance gesture and audition and the PC Set theory idiomatically applied to the guitar.
- Computer modeling of the framework.
- Open source tool for interacting with the framework.

2 Conceptualization and Related Work

2.1 Comprehensive Approach to Improvisation

In this work, the approach adopted acknowledges improvisation as discussed by authors like Andrew Goldman (2020) [4] and Jonathan Leathwood (2019–2020) [7,8]. According to their perspectives, improvisation serves not only as a means to introduce novelty and real-time solo creation in music but also as a valuable tool for learning and practicing technical musical concepts, repertoire exploration, and enhancing comprehension of diverse stylistic frameworks. Furthermore, improvisation is recognized as a catalyst for musical creation, drawing upon various elements found within the Western classical music tradition.

2.2 Classical Guitar Improvisation

The practice of improvisation has played a significant role within the tradition of classical guitar, spanning from the Baroque period to the classical and romantic eras. During this time, improvisation was highly regarded in the context of instruments like the Spanish vihuela and baroque guitar, with notable figures such as Gaspar Sanz and Francesco Corbetta showcasing their techniques and ornamentation practices. As musical styles evolved, there was a notable transition from ornamentation-based improvisation within composed works to more extensive improvised sections based on the principle of variation within guitar compositions. Influential guitarists and composers like Fernando Sor and Mauro Giuliani made noteworthy contributions to improvisation techniques, impacting the role of the guitarist as both performer and composer.

However, the practice of classical guitar improvisation experienced a decline during the late nineteenth and early twenty centuries, mainly due to the increasing focus on notation-based performance practices. Nevertheless, there have been notable revivals of interest in classical guitar improvisation in the late twenty and twenty-first centuries. These revivals were spearheaded by innovative practitioners and educators, such as Roland Dyens and Dusan Bogdanovic, who drew inspiration from cross-cultural influences, including jazz, world music, and experimental avant-garde movements. These practitioners embraced improvisation and integrated it into formal education, exploring novel teaching approaches.

These developments have revitalized the practice of classical guitar improvisation, encouraging its inclusion in contemporary performance and pedagogical contexts. By incorporating diverse influences and innovative approaches, improvisation has evolved to meet the demands of the modern musical landscape while retaining its historical significance within the classical guitar tradition.

2.3   Classical Contemporary, Post Tonal Music, and Set Theory

Classical contemporary music encompasses a wide range of musical styles and techniques that emerged in the twenty and twenty-first centuries, challenging conventional notions of melody, rhythm, and form.

Classical contemporary music is an art form of the present day that actively engages with the inherited tradition of Western art music while also seeking to push boundaries and explore new sonic territories. Furthermore, classical contemporary music is characterized by its pluralism of styles and aesthetic approaches, reflecting a diversity of individual compositional voices. Composers within this genre often draw inspiration from a variety of sources, including other art forms, cultural traditions, and contemporary social and political issues.

Within Classical contemporary music and according to musicologist Richard Taruskin (2010), post-tonal music is defined as "music that avoids clear tonal centers and traditional tonal hierarchies" (Taruskin, R 2010. p. 413) [11]. This compositional approach seeks to expand the harmonic vocabulary by utilizing dissonance, atonality, and various pitch organization systems such as serialism, Set theory, and pitch-class analysis.

Moreover, music theorist and composer Allen Forte offers a more technical definition, describing post-tonal music as "music in which tonality and the hierarchical relations of pitch are not primary determinants of the musical structure" (Forte. 1973. p. 4) [3]. This definition highlights the shift away from tonal hierarchies and the exploration of alternative methods for organizing musical material. Initially proposed by Forte himself, Set theory offers a formalized approach to understanding the organization, relationships, and transformations of pitch materials in compositions, contributing to a deeper analysis and interpretation of musical structures. Set theory also introduces operations such as union, intersection, complement, and transformational operations like transposition and inversion. These operations enable the examination of musical relationships, symmetries, and transformations within and between pitch-class sets. While Set theory has been influential in music theory, it is worth noting that its applications and interpretations can vary among scholars and composers.

2.4   Digital Tools for Set Theory Analysis

Digital tools have revolutionized the application of set music theory [13], offering musicians and scholars resources for analyzing and manipulating pitch-class sets. Here is an overview of some prominent digital tools used in the context of set music theory:

- PC Set Calculator (by David Walters) [14]
- Music Theory Set-Class Calculator (by Jeremiah Goyette) [5]
- Set Calculator (by Dan Román) [9]
- Set Theory Calculator (by Jay Tomlin) [12]

Nevertheless, it is worth noting that while these tools are intended for extracting and visualizing information to facilitate musical analysis, none of them are specifically designed to offer a framework for creative purposes or cater to the specific needs of guitarists.

3 Guitar Improvisation Harmonic Framework Based on Set Theory and the Exploration of the Mechanical Gesture

Constructing a framework for achieving harmonic coherence in improvisation through the exploration of the mechanical gesture involves two key aspects. Firstly, it involves finding ways to produce different sonic outcomes by developing and evolving a particular mechanical gesture. Secondly, to develop the proficiency to establish distinct structural relationships among these sonic outcomes. In order to achieve this end, we will now introduce and explain our harmonic perspective grounded in Set theory. Then, we will outline our approach to exploring and evolving the mechanical gesture as a seed for generating improvisational materials. Subsequently, we will explore the relevance of proposing a digitally assisted approach to implement this framework.

3.1 Harmonic Perspective: From an Analytical to a Creative Perspective of Set Theory

One key aspect of set theory is the concept of pitch-class sets as equivalence classes. It recognizes that different transpositions and inversions of a pitch-class set share essential musical characteristics and can be treated as the same set. This allows for a more abstract analysis of musical structures, focusing on the relationships between pitch classes rather than specific pitches. Guided by this simple principle, and going from an analytical to a creative point of view, we can state that moving from one set to another can be perceived in a way that one can determine relationships of unity, transition, or contrast because of the harmonic content that underlies on each set.

In this context, the concept of the IC Vector, another pivotal aspect of set theory, becomes instrumental in providing us with a harmonic perspective to establish relations of unity, transition, and contrast. Serving as a representation, the IC vector furnishes valuable information on the intervallic content of a set, focusing on interval classes. Each IC captures the frequency of occurrence for every possible interval class within that set. Analyzing the degree of similarity between IC vectors from different sets of notes facilitates the determination of harmonic equivalences, identification of shared characteristics or distinctions between sets, and exploration of opportunities for unity, transition, and contrast within a musical discourse.

According to Joseph Straus (1991) [10] the achievement of compositional unity and diversity through the selection of a pitch-class set as a fundamental structural unit from an analytical perspective. However, our proposal expands on Straus's observation that coherence is established through the use of similar pitch-class sets. We argue that harmonic consistency can be further attained by examining the evolution of the interval class, as manifested in the IC vectors.

This implies, of course, understanding the notion of a set and its interval content as a primary unit or seed of the audition and meaning in musical discourse, rather than establishing hierarchical relationships departing from the pitch of a single note as a center of gravity as it is in the tonal system. However, this approach does not implies, as it is for instance in twelve-tone serialism, to avoid tonal implications, but allows one to freely enter or leave episodes that can be perceived as tonal in a succession of sets, enriching the overall harmonic frame of a piece.

3.2   Exploring the Mechanical Gesture

On the other hand, and for the purpose of designing a harmonic framework that is linked to the exploration of the guitarists' mechanical gesture, it is necessary to inquire about the way in which mechanical memory is linked to audition and performance, and even more so for our purposes with improvisation.

Muscle memory allows guitarists to automate physical movements, enabling them to concentrate on musical expression and interpretation. In other words, to make playing possible, performers develop an interdependence between mechanical operations and cognitive processes, such as attention allocation and musical decision-making. In the case of classical guitarists, this is concreted in mechanical gestures that require coordination between both hands and fingers, including precise placement, as well as the execution of complex chord voicings, single melodic lines, polyphonic textures and patterns, and techniques such as arpeggios, scales, and various types of strumming. As discussed in the introduction, in classical guitar tradition, the dominant approach to performance emphasizes the coordination between mechanic operations and auditory memory, following the line of memorizing repertoires and internalizing musical imagery, synchronizing the physical actions with the internalized auditory representations to enhance accuracy, expressiveness, phrasing, and musical communication. This frequently results in the fact that, for a performer, one complex mechanical gesture corresponds to one concrete auditory and musical representation in a concrete piece of repertoire. This frequently limits the contact with improvisation since, for this practice, it is best to approach and exercise the study having different musical outcomes for similar mechanical gestures, this, is to develop adaptability, and the ability to generate novel musical ideas in real-time performance and skills to focus in the link between audition and mechanical operations.

To accomplish this and adhere to our harmonic perspective, the following principles need to be established:

1. The same mechanical gesture can produce different sets of notes. In other words, performing a specific mechanical fingering in various regions of the guitar, be it the left or right hand, will yield different note sets while retaining the same gestural character.

2. We can describe the similarity between two fingerings by introducing the concept of mechanical modulation. This means a mechanical gesture can be adjusted to create other gestures. For instance, maintaining a finger's position in the right hand while introducing some variation in the left hand's finger design. Therefore, the degree of similarity between gestures will depend on the extent of modulation applied.
3. Introducing the notion of mechanical modulation allows us to consider and observe the development and evolution of a mechanical gesture within a musical context.
4. To achieve structural coherence in a musical discourse derived from the evolution of a mechanical gesture, it is essential to integrate it with a harmonic perspective.

As we will explore further, integrating the domains of harmonic perspective and mechanical exploration requires carrying out several tasks and operations that demand advanced theoretical knowledge and guitar skills. This process can be time-consuming and labor-intensive, but it could be significantly simplified through the use of a digitally assisted approach.

3.3   Towards a Digital-Assisted Approach

In order to unify in one framework our vision of developing an improvisation approach based on Set theory and our interest in integrating the exploration of the mechanical gesture, it becomes essential to delineate the required operations that performers need to undertake in both the harmonic and mechanical domains.

First, following our line of approach and the principle of establishing relationships that encompass unity, transition, and contrast among various sets of notes based on their intervallic content as informed by Set theory, performers would engage in a technical endeavor that involves executing the following calculations and processes:

1. Choose a fundamental set that acts as a starting point for generating the harmonic discourse.
2. Perform transposition of the individual pitches within the selected seed set.
3. Identify the fundamental harmonic relationships present within the set.
4. Quantify the number of interval classes within the set to precisely determine the hierarchy of intervallic content, resulting in the creation of an IC vector.
5. Explore and generate additional sets that share the same IC vector, aiming to establish harmonic equivalences within a set group. This process aligns with our approach, ultimately leading to the creation of a diverse harmonic palette comprising multiple sets that exhibit a coherent sense of harmonic unity.
6. Engage in experimentation by exploring different possibilities for rearranging the resulting sets through various dispositions and inversions.
7. Skillfully modulate the harmonic content by altering the IC vector of different sets, leading to the creation of a new palette that facilitates the establishment of a sense of transition and contrast when performing these sets.

8. Iterate step 6 with the modulated sets to develop sections of transition and contrast throughout the musical discourse.

On the other hand, when it comes to exploring the mechanical gesture, guitarists have two paths to explore within this domain. The first approach involves modulating a mechanical gesture to create alternative variations and then analyzing the potential resulting sets to determine whether they can be utilized for developing a discourse that demonstrates coherence and mastery over its harmonic development. In a structured context rooted in Set theory, this would involve examining the interval content of the various configured sets. Within the execution of each mechanical gesture, one would determine its IC vectors and then ascertain whether there exists a sense of unity, transition, or contrast between the sets by comparing their IC vectors. Through this iterative process, a musical discourse can be developed. The second approach involves analyzing the set formed during the execution of a specific mechanical gesture, following the eight steps proposed earlier for the harmonic method based on Set theory. Subsequently, the guitarist would search for fingerings on the guitar that enable them to align the mechanical gesture in accordance with the logic of harmonic discourse.

As it becomes evident, the process of addressing both the harmonic and mechanical domains involves a complex series of steps and operations, requiring significant time and effort from performers to acquire the necessary skills to become fluent in the new language. To overcome this challenge, we propose the implementation of a digital assistance solution that can streamline these operations and simplify the preparation and practice of improvisation within our proposed framework. By integrating such a solution, we can unlock the advantages of an improvisational approach that enables exploration of the languages of contemporary classical music within the idiomatic context of the guitar, firmly rooted in the fundamental exploration of the mechanical gesture.

4   Computer Modeling of the Framework

In order to make this approach feasible, it is fundamental to leverage computing power to compute the myriad of calculations required for obtaining the requirements for analyzing both; mechanical modulations of a given gesture and harmonic content of the sets by getting their IC Vectors. The first step towards this goal is to model the technique in a computational way. For this, we need to first **model the overall problem** as an input-processing-output system, and then model the **inputs** and **outputs** in a way understandable to the musical audience. The rest of this section describes three modelings, that were later used to implement an interactive system. We present the modeling of the problem as a contribution to this work, while the details of the interactive prototype are going to be presented in future work.

4.1   Overall Problem Modeling

To manage the scope of the framework, we choose to focus on the specific task of helping guitarists generate and explore possibles sets of notes, that are based on an inspirational mechanical gesture (seed), which are feasible for the guitar, and which follow the proposed Framework based on the Set theory idiomatically applied to guitar.

As a guiding example, let's think of Alicia, a guitarist who has been studying Leo Brouwer's Etude 6, and has found inspiration in a gesture that naturally implies a mechanical gesture and a set of notes. She wants to improvise a piece of music inspired by that. The framework should then generate all the possible solutions that adhere to a group of restrictions. The main restrictions are **mechanical, angular**, and **harmonic**. The mechanical and angular restrictions are inherent to the guitar, while the harmonic pertain to the Set theory. The framework then generates a series of solutions, that have harmonic **unity**, this is the same IC vector, while also having the same mechanical and angular style as the seed. She learns to practice these but then wants to add more flavor to her piece, so she starts exploring other solutions that would generate **contrast** on her improvisation, so she **modulates** the harmonic restriction. Then, to make everything play together nicely, she generates passages that would be in between to have some **transition** on her musical discourse. Finally, and because she is a skilled guitarist she explores modulating also the mechanical and angular restrictions for even more sets that she can add to her repertoire. Then, having all of these solutions, she can practice them for an improvisation concert or maybe compose a piece.

The whole modeling of the framework for this example is summarized in Fig. 1. The model starts with a seed, which is a mechanical gesture that produces a set of notes that the guitarist wants to use as inspiration. This is then passed by an option generator, which will have an algorithm to compute all the possible solutions (other sets of notes) that will satisfy the mechanical, angular, and harmonic restrictions. Then, the guitarist can modulate the options for each one of those three restrictions, to generate unity, transition, and contrast based on their similarity to the original seed. To compute the similarity, we propose distance metrics for the mechanical, angular, and harmonic restrictions.

4.2   Modeling Solutions and Inputs

In the framework, solutions are modeled as an array of pairs of integers, representing the combination of fret and string that corresponds to the fingering represented. As an example, fourth fret string two is represented as $[2, 4]$, and the full fingering shown in Fig. 2 is modeled as $[[1, 0], [5, 0], [2, 4], [3, 6], [4, 7]]$.

Distances. In order to model the similarity, we defined distance functions for mechanical, angular, and harmonic restrictions.

$$\Delta_{mec}(A, B) = \Sigma_{i=0}^{n}[(Fret_i(B) - Fret_i(A))^2 + (String_i(B) - String_i(A))^2]$$

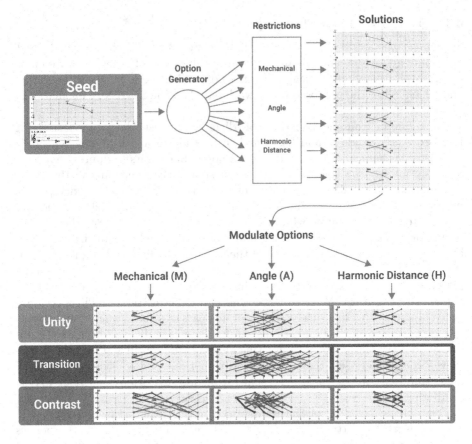

Fig. 1. Guitar Improvisation Harmonic Framework problem modeling

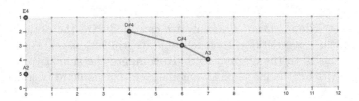

Fig. 2. Example of a fingering and how it is represented and modeled in the framework

Where a and b are the set notes to compare, and i is the note position in the set starting at 0. Basically, this is an Euclidean distance in the Cartesian plane of the Frets and Strings, but ignoring the square root to speed up the computation as shown in Fig. 3.

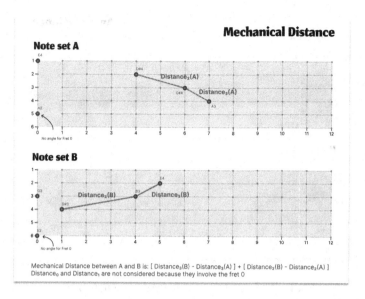

Fig. 3. Distance between two note sets

The **angle distance** is expressed by computing the difference of each one of the independent angles in the set

$$\Delta_{ang}(A, B) = \Sigma_{i=0}^{n-1}[(Angle_i(B) - Angle_i(A)]$$

Where Angle is computed using the following formula, and only on notes that aren't on the fret 0 (although there is an option to include these in the calculation)

$$Angle_i(A) = tan^{-1}\left(\frac{Fret_{i+1}(A) - Fret_i(A))}{String_{i+1}(A) - String_i(A)}\right)$$

With an extra option to ignore the sign of the angle, which produces solutions with mirrored angles. Figure 4 visualizes how the angle distance would be calculated between two note sets.

The Harmonic distance is computed by calculating the element wise distance between IC Vectors

$$\Delta_{harm}(A, B) = \Sigma_{j=0}^{6}[(ICVector_j(B) - ICVector_j(A)]$$

Where j is the position in the ICVector, which has six positions in total.

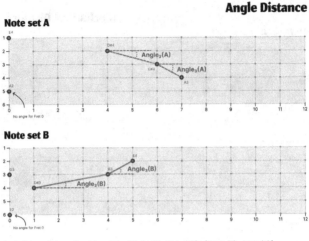

Angle distance between B and A would be: [Angle₃(B) - Angle₃(A)] + [Angle₂(B) - Angle₂(A)]

Fig. 4. Angle Distance between two note sets

4.3 Modeling Harmonic Perspective

The framework poses the concept of harmonic perspective, aiming to establish categories of unity, transition, and contrast based on the level of intervallic correspondence between sets. This entails assessing the degree of similarity between their IC vectors, thereby reflecting the intervallic content of the sets. As shown in Fig. 5 To capture the concepts of Unity, Transition, and Contrast, for each one of the distances, we set the following ranges:

	Unity	**Transition**	**Contrast**
Harmonic Distance Δ_{harm}	$[0, 3]$	$(3, 10]$	$(10, \infty)$

Harmonics perspective

Similar		Opposite
Unity	Transition	Contrast
Same IC Vector		Different IC Vector

Fig. 5. Guitar improviser's Harmonic perspective

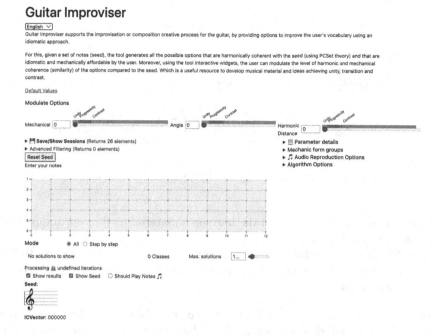

Fig. 6. Guitar improviser interface

5   Guitar Improviser: Preparation and Practice Digital Tool

Guitar Improviser is a digital open-source tool developed in accordance with the framework we have described. It seamlessly integrates our approach, combining the exploration of the guitar's mechanical gesture with the inherent restrictions and possibilities it offers, along with our utilization of Set theory to establish a harmonic context that enables the formation of relationships involving unity, transition, and contrast. The tool is designed for guitarists by providing them with the opportunity to experiment with various possibilities for creating a coherent musical discourse. Guitar Improviser interface as shown in Fig. 6 and can be accessed on https://infovis.dev/viz/guitarImproviser.

6   Example of Basic Use

In this section, we present the basic use of the Guitar Improviser tool to prepare and execute an improvisation based on the Etude 6 for guitar by Cuban composer Leo Brouwer. This sample embodies the framework and concepts discussed in this paper. To begin, we extracted the set of notes generated by the opening gesture depicted in the Etude as shown in Fig. 7.

To start using the Guitar improviser tool, we proceed to introduce the left hand fingering by clicking on the fingerboard shown on the interface, this gesture

Fig. 7. Opening gesture Etude 6 by Leo Brouwer. 1978. Editions Max Eschig

will be used as the seed to obtain the harmonic material used in our improvised piece as shown in Fig. 8.

The **Modulate Options** are sliders that allow the user to manipulate the parameters that configure both dimensions discussed in our conceptual and computer modeling approach:

1. Harmonic: Meaning that moving the **Harmonic Distance** slider allows the user to obtain solutions of remaining or changing the IC vector, thus modulating the intervallic content of the seed.
2. Mechanical gesture evolution: Meaning that moving the slider allows the user to manipulate the two parameters (**Mechanical and Angular**) that configure the mechanical gesture evolution.

Modulating options allows one to configure particular settings. Guitar improviser also displays information about the seed. This includes score notation as Pitch classes, this is, that it does not consider octaves. The basic use of the tool also shows the number of solutions calculated by a given setting.

By clicking on the **Show results** and **Show Seed**, and moving the **Max. solutions** slide bar the tool shows solutions for a given setting as shown in Fig. 9. This particular setting having the **Modulate Options** in 0, shows solutions that have perfect coincidence with the seed in both Harmonic and Mechanical dimensions. This means that all solutions have the same IC vector, thus the same interval content, and, that they also exhibit the same fingering properties of the seed in terms of distance and displayed angles between fingers.

Step by step check circle allows one to visualize every single solution as shown in Fig. 10 and Fig. 11.

By modulating the **Harmonic Distance** slide bar, the user can obtain solutions that vary the level of harmonic coincidence compared with the seed. Following our conceptual approach means that by doing this, one can achieve motions of unity, transition, or contrast through the musical discourse, depending on the overall harmonic frame of the piece. See Fig. 12.

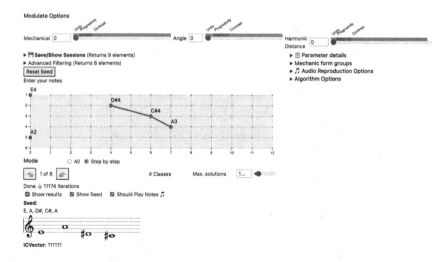

Fig. 8. Seed generated by opening gesture on measure 1 of Etude 6 by Leo Brouwer

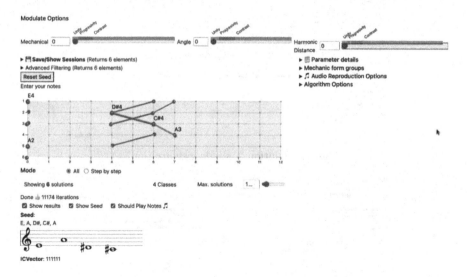

Fig. 9. Opening gesture with **Modulate Options** in 0

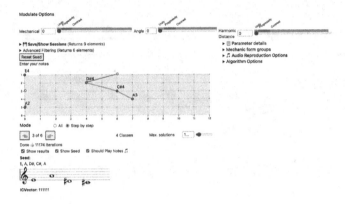

Fig. 10. Solution 1. single visualization

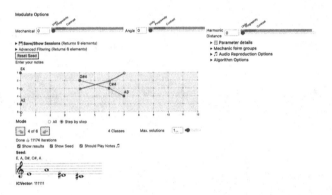

Fig. 11. Solution 2. single visualization

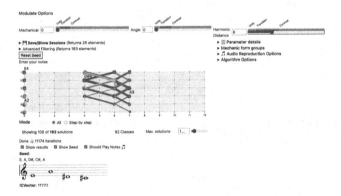

Fig. 12. Harmonic Distance slide bar use

On the other hand **Mechanical** and **Angle** slide bars allow the user to modulate the mechanical gesture and therefore to reinforce the sense of unity, transition, or contrast by reaching registers close by or far from the seed as shown in Fig. 13 and Fig. 14.

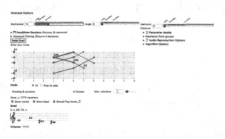

Fig. 13. Mechanical slide bar use

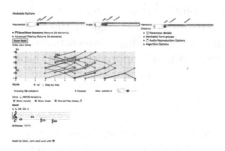

Fig. 14. Angle slide bar use

In the video at https://infovis.dev/viz/guitarImproviser/Guitar_Improviser_ demo_small.mp4 from minute 0 to 3:43, we detail the illustration of the basic use using the seed taken from the initial gesture of the Etude 6 by Brouwer. From minute 3:44 to the end, a piece is improvised using the opening gesture of the Etude and following our framework assisted by the Guitar improviser tool.

7   Conclusions

Carrying out an approach to improvisation on contemporary classical guitar that allows to navigate its wide stylistic spectrum and that starts from the exploration of the evolution of the mechanical gesture, requires integrating this experimentation with a framework of structural coherence.

The conceptual framework we present necessitates digital assistance due to the intricacy of the operations a performer must undertake to carry out the calculations involved in evolving a mechanical gesture and controlling the aspects of unity, transition, and contrast that we propose as essential categories in the development of musical discourse.

This digital assistance empowers the performer to discover solutions provided by the tool, which seamlessly integrates the control of mechanical gesture evolution and harmonic expression. As a result, the performers can liberate themselves from these complex calculations and focus on developing the mechanical gesture in a more intuitive manner, guided by their sense of hearing. This approach aligns with other musical traditions that rely heavily on the auditory sense and the oral tradition. Hence, the framework we introduce, while exhibiting stylistic compatibility with the context of classical contemporary music in terms of its sound outcomes, also integrates approaches from other traditions when it comes to approaches to the process of thinking about improvisation and musical creation.

Currently, the Open source tool we have created enables us to approach improvisation preparation. However, to operate it in real-time live performance, we would need to adapt the interface to facilitate user interaction without using the mouse.

Composers may also find value in considering the digitally assisted framework we present for their own compositions. By configuring it through the exploration of mechanical gestures, this framework ensures the technical and idiomatic feasibility for guitar, while also enabling composers to experiment with diverse harmonic contexts grounded in our approach to set theory.

8 Future Work

Future endeavors encompass both conceptual and practical aspects, enabling the incorporation of additional frameworks to explore the development of various parameters inherent to improvisation. These parameters encompass the evolution of rhythm, timbre, dynamic ranges, articulation, and musical form.

On the other hand, as has been said, projecting this framework beyond the preparation of improvisation to take it to the moment of live performance requires adapting the interface and integrating other devices such as tablets and page-turner pedals.

Future work in this project also includes new papers and presentations that delve into the development of the code that Guitar improviser works with, as well as new user guides and audiovisuals that illustrate the advanced use of the tool.

Finally, we want to emphasize that Guitar improviser's prototype and interface showcased in this paper serve a demonstrative purpose, illustrating our proposal of a digitally assisted framework for studying and preparing improvisation and creation on the contemporary classical guitar ambit along with its logical and computational modeling. However, it's important to note that the development of the algorithm, the code, and the advanced operation of the prototype constitute other contributions that will be fully detailed and addressed in a future paper, where we will elaborate on and acknowledge these specific aspects.

References

1. Caporaletti, V.: An audiotactile musicology. RJMA-J. Jazz Audiotactile Musics Stud., 1–17 (2018). https://www.academia.edu/37643283/An_Audiotactile_Musicology?email_work_card=view-paper

2. Després, J.P., Burnard, P., Dubé, F., Stévance, S.: Expert western classical music improvisers' strategies. Expert western classical music improvisers' strategies. J. Res. Music Educ. **65**(2), 139–162 (2017). https://doi.org/10.1177/0022429417710777

3. Forte, A.: The Structure of Atonal Music. Yale University Press (1973). https://www.google.com/books/edition/The_Structure_of_Atonal_Music/j9aV2JYHY4AC?gbpv=1

4. Goldman, A.J.: Improvisation as a way of knowing **22**(4) (2016). https://mtosmt.org/issues/mto.16.22.4/mto.16.22.4.goldman.html

5. Goyette, J.: Music Theory Set-Class Calculator. https://jeremiahgoyette.com/calc/set_class/

6. Hogan, N.: Why don't classical musicians improvise? (2022). https://songbirdmusicacademy.com/why-dont-classical-musicians-improvise/

7. Leathwood, J.: Improvising changes, Part 3: the hexachord, 6–14 (2019). https://www.youtube.com/watch?v=ba-J1LzeYwI&t=1s

8. Leathwood, J.: Improvisation as a Way of Knowing: Baroque (2020). https://www.youtube.com/watch?v=rZJFt6SuHJM

9. Román, D., Solomon, L.: Set Calculator. http://foreversound.com/setcalculator/main.asp

10. Straus, J.N.: A primer for atonal set theory. Technical report (1991). https://www.jstor.org/stable/40374122

11. Taruskin, R.: Music in the Late Twentieth Century: The Oxford History of Western Music. Oxford University Press (2006). https://books.google.com.co/books?id=mWFmDwAAQBAJ&printsec=frontcover&source=gbs_atb#v=onepage&q&f=false

12. Tomlin, J.: Set Theory Calculator (2022). http://www.jaytomlin.com/music/settheory/

13. Tucker, G.: A Brief Introduction to Pitch-Class Set Analysis (2001). https://www.mta.ca/pc-set/pc-set_new

14. Walters, D.: PC Set Calculator (2001). https://www.mta.ca/pc-set/calculator/pc_calculate.html

Dignitas in the Metaverse

Savithri Bartlett[⊠], Sam Chester, Philip Delamore, Sebastian Roeck,
and Zowie Broach

Royal College of Art, Kensington Gore, London SW7 2EU, UK
S.Bartlett@rca.ac.uk

Abstract. In response to increasing calls to address "human dignity" in the
Metaverse, the Royal College of Art's MA Fashion Programme utilised Epic's
"Metahuman Creator" software to create multiple digital identities. Human dignity has its etymological root in the Latin 'dignitas', and is translated as worth. For
the purpose of this study, its meaning is derived from the Charter of the United
Nations (1945) which defines the fundamental rights of the humacy3n person,
their equality, dignity and worth.

Epic's "Metahuman Creator" is a new type of digital human configurator,
based on pre-existing scans of real people where only physically plausible adjustments can be made. As it has a vast range of facial features and skin tones the user
can design a variety of digital characters. However, its tools representing trans identities and invisible disabilities have limitations. Here we show the experience of
two users, one with a hearing disability and the other undergoing gender-affirming
therapy.

We found that tools for representing their identity fell short of expressing their
inherent human dignity. The foundations of the Metaverse are therefore at risk
of replicating inaccessibility and excluding persons with disabilities and gender
differences.

We anticipate our paper to be the starting point for wider participatory research
that involves all user groups in the design of platforms, tools and systems. We
foresee the findings of our paper fulfilling the UN Charter on Human rights in
the digital space through the shaping of policies and advancing existing standards
(ISO).

Keywords: Metahuman · Human Dignity · Disablement · Disablism · Embodied
Human Diversity · Agency · Creativity and Design · Digital Bodies · Gendered
Bodies · Inclusion

1 Introduction

1.1 Background and Context

The Metaverse. Since the new millennium, advances in Information and Communication Technologies (ICT) such as real-time game engines, computer graphics processing,
rendering (ray tracing) and 5G broadband speeds, has lead to a rise in Massive Multiplayer Online Games (MMPORGs), as well as open-world games that rely on User

© ICST Institute for Computer Sciences, Social Informatics and Telecommunications Engineering 2024
Published by Springer Nature Switzerland AG 2024. All Rights Reserved
A. L. Brooks (Ed.): ArtsIT 2023, LNICST 564, pp. 186–201, 2024.
https://doi.org/10.1007/978-3-031-55319-6_13

Generated Content. More recently, the global pandemic has accelerated digitalisation across all sectors with the promise of an emergent "Metaverse", a term coined by Neal Stephenson in his science fiction novel Snow Crash (1992) to describe a single, shared, persistent, 3D virtual space, inhabited by Avatars converging virtual reality (VR) and the internet.

The Metaverse, as defined by Ball (2021:42), has an infinite number of users "with an individual sense of presence, and continuity of data" such as gender, race, transactions, ownership and communication. Whilst opportunities are associated with the Metaverse, it is at an embryonic stage and how it develops is open to conjecture. As Floridi (2022:09) observes, it is "a lot of science fiction, little technology, and even less understanding of human nature".

Role of Avatars and Digital Embodiment. Avatars will play a pivotal role in the Metaverse, enabling users to transcend geographical boundaries, interact socially and embody, i.e., express meaning, experience, and agency. Digital embodiment, according to Vasalou et al., (2008) allows users to (i) accurately reflect their offline selves by displaying stable self-attributes (ii) construct a playful representation of the self and (iii) send an embodied message. The contrast between the physical body of the user and the digital body of the avatar is described by Schultze (2011) as a form of "technological artefact". The limitations of software tools and interfaces to allow or disallow certain types of bodies to exist has been identified by Harper (2020) as "gaps and biases" that become the cultural norms.

In this paper, we refer to marginalised individuals who are identified as "transgender" or "disabled", i.e. those whose gender identity is incongruent with the sex assigned to them at birth, or who have been described as outside of able-bodied norms. Research by van Aller (2018) showed that gender fluid and trans participants felt limited by the lack of options to create non-normative bodies in games. Morgan et. al., (2020) studied adolescent trans and gender diverse (TGD) youth who reported negative experiences of avatar design. TGD gamers recommended subtle design changes to not just the visual appearance but also to the use of language by providing "customisable pronouns" that would not limit them to male or female in-game interactions. Kosceisza (2023) studied non-cis gender participants who engaged in varying levels of participation with other players via voice chat. One, who was medically transitioning from female to male appreciated the moment their voice (deepened through the use of hormone treatment) aligned with their avatar's appearance and the "heroic and deep, chocolate sounding" masculine tones. In voice interactions with other players, they felt further affirmed as male by the wider gaming community. This was in contrast to other players who identified as nonbinary or trans-feminine who made "strategic choices" as to whether they should speak in chats with unknown players due to prior experience of abuse and misgendering, when their voice was perceived as a mismatch with their female avatars.

Research by Zallio & Clarkson (2022), Radanliev et. al. (2023), and Seigneur et al. (2023) investigated accessibility issues for disabled users in the Metaverse. They found that in spite of the proliferation of Metaverse projects, access to the virtual world "remains distant" for those with disabilities. The authors propose the use of multiple emerging technologies (VR, AR and IoT), to better enable physically disabled participants to engage in the design and development of standards for the Metaverse.

The current study stands apart from the above due to its focus on a digital human configurator, and the involvement of participants, one with a hearing impairment and the other at the start of gender-reassignment therapy. Digital human configurators, such as the Metahuman Creator software, are different from traditional gaming character customisation and avatar generation tools. It is a new genre of digital technology that models and textures photorealistic human identities, allowing users to configure a "Metaself" in a plausible but data-constrained way, outside of any pre- existing games and virtual worlds.

Project Talks and Workshops. The Metahuman project began with a series of four workshops to unlock creativity and innovative future-led design thinking. It was led by philosophers, social scientists, evolutionary biologists, academics, artists and curators who presented topics ranging from gene editing to psychological and legal definitions of identity, biometric surveillance and the curatorial practice of queering.

In a workshop on human dignity, the Transhumanist philosopher, Dr Stefan Sorgner argued that the greatest potential revolutionising our way of life can be found at the intersection of digitalisation and precision medicine and gene editing technologies, "so in addition to the outer Internet of Things (IoT), we develop an Internet of Bodies (IoB), that are devices which alter the body's functions." The IoB ecosystem, however, is a largely unregulated market that poses cyber and privacy risks to the uniquely sensitive personal data that devices gather on humans. Professor Jannice Käll, a specialist in Posthuman law and digitalisation, discussed the risks associated with the blurring of boundaries between physical and digital bodies, which might lead to our multiple identities being owned by or "streamlined towards the ends of capitalist or racist corporations."

The workshops challenged participants to question:

1. How do we become Metahumans?
2. What do we want to take with us into the Metaverse, and what do we want to leave behind?
3. What might the effect on our physical identity be of creating a new digital identity for the Metaverse, and to what extent might our personal and social identities of ability, gender, race, sexuality, age and class be redefined in the Metaverse?

2   Methodology

We conducted a series of focus groups and interviews with 17 participants across 2 months. Participants voluntarily enrolled on the project from across the School of Design at the Royal College of Art, ranging in subject specialism from Industrial Design Engineering to Intelligent Mobility, Design Products, Textiles and Fashion. Participants had an average age range of 23, with 7 participants identifying as male, 9 as female and 1 as trans. When it came to race, 3 participants identified as White, 12 participants were Asian, 1 was Hispanic and 1 was of British-African origin. All participants were asked to sign a consent form prior to engaging in the study.

During these individual and small group sessions, participants described their design process and use of facial recognition animation from Livelink and audio-driven facial animation from Speech Graphics that drove not just the mouth but the entire face of their avatars. The term "avatar" is taken from Sanskrit to mean the human embodiment of a deity that interacts with humans and experiences Earth from a human perspective (Nowak, 2015). For this study, the term 'avatar', 'Metahuman', 'digital bodies' are taken to mean "the user in a fully immersive digital environment" where it can "facilitate complex actions including nonverbal communication via gestures, body posture, proxemics and even haptics" (Nowak and Fox 2018: 33).

The work of two participants will be shared in case study format and narrated in first-person, singular. These case studies focus on human dignity, trans identities and disability in the virtual world. The participants' experience of the Metahuman Creator software and other relevant digital technologies are cited in Table 1.

Table 1. Participants' experience of the Metahuman Creator software and other relevant Digital Technologies

	Metahuman Creator	Mesh to Metahuman	Agisoft Metashape (standard edition)	Artec Eva Colour Scanner	Unreal Engine	Blender	Speech Graphics	Live Link Face
Sam	!	!	!	!	!	!	!	!
Seb	!				!		!	
Pros	-Free -Minimal 3D experience required -Browser based -High fidelity -Ready for animation -Interoperable with Unreal Engine + Live Link	-Free -Users has full control over uploaded 3D scan -Possible to create a Methuman with much closer likeness to user	-Industry quality photogrammetry software -User has full control over input - possible to photograph and upload a diverse range of bodies/faces	-High quality capture resolution -Handheld and easy to move around a body -User has full control over input - possible to photograph and upload a diverse range of bodies/faces -Faster and more accurate than photogrammetry in capturing form	-Free -Huge library of free 3D models	-Open Source -Freedom to create any 3D object	-Automated facial animation based on voice recording	-Realtime facial/head animation -Makes facial tracking more accessible to beginners
Cons	-Constrained within preset "physically plausible" sliders -Minimal choice of body types, hair types, makeup, clothing	-User must generate their own 3D mesh -"Real" textures from 3D scan not preserved -Mesh can be "blended" with preexisting Metahuman scans to remove errors/artefacts -No ability to use a body scan so face is applied to a limited choice of presets	-Cost -High level of skill required for accurate results -Lighting conditions to capture photos can be difficult to get right -When capturing a person, they must remain as still as possible until all photos are taken -Mesh generated must be taken into another software to prepare for animating	-Cost -High level of skill required for accurate results -Mesh generated must be taken into another software to prepare for animating	-High level of skill required -Minimum System Requirements -Poor interoperability outside Epic softwares -Unable to build avatars within Unreal must be exported from elsewhere	-High level of skill required -Minimum System Requirements	-Cost -Creates generic facial animations -Lacks personality	-Requires Iphone 12 (or above) -Unable to detect very subtle facial movements -Requires knowledge of Unreal Engine to process and render data

(*continued*)

Table 1. (*continued*)

Metahuman Creator	Mesh to Metahuman	Agisoft Metashape (standard edition)	Artec Eva Colour Scanner	Unreal Engine	Blender	Speech Graphics	Live Link Face
-unlabelled but clearly gender marked							
binary options of male and female							
(facial features + body types)							
-Unable to create a body with any							
form of disability							

3   Results

3.1   Portraying Gender Non-conformity in the Metahuman

Case Study #1: Sam. I am a digital artist and craftsperson whose work orbits ideas of trans identities and storytelling, moving between the real and unreal, the physical and digital. I began my medical transition a year ago at the age of 29. This came after years of questioning my gender identity and struggling with dysphoria about how others perceived me. My earliest memory of feeling disconnected from the gender I was assigned (male) was at the age of 7. I first understood about gender fluidity when I moved to London at 19 and began identifying as non-binary but with an emphasis, I felt more feminine than masculine. In the 10 years since I have felt my identity evolve as I have come to better understand myself and feel most comfortable navigating the world.

I began to experiment with representing myself in the digital space whilst I was a student on my undergraduate Fashion degree at Central Saint Martins (CSM). I experimented with 3D photogrammetry and photographed myself in 6 different garments (Fig. 4). These existed as static models capturing a moment in time and lacking any movement. When I embarked on my Masters study at the RCA in 2019, I was able to get a complete body scan using an Artec Eva colour scanner. It provided me with a higher resolution 3D model (Fig. 5) than I had achieved previously. As I became more experienced with 3D modelling and animation techniques I was able to customise my scan to represent my idealised body. After graduation, I continued at the RCA as an Associate Lecturer and honed my practice. I joined the Metahuman Project as a technical adviser. Working with Metahuman Creator I was able to create a digital human with a much higher fidelity, elevating my work to industry standard. With the initial Metahuman I used tools such as Live Link Face and Speech Graphics to animate my avatar, a task that would usually take several highly skilled individuals familiar with various 3D techniques to accomplish.

Aim. The aim of my Metahuman was to try and recreate my trans identity in a virtual avatar. I have always been drawn to virtual spaces as a way to exist in a non-gendered body, a body made of 0's and 1's, free from the constraints of the physical world.

Fig. 1. Screenshots of the Metahuman Creator with PureRef open (left). (source: Author's own)

Creative Process. My initial experience of the Metahuman Creator was of an awkward software that was slow to respond. The only way to create something remotely like me was to either sit in front of a mirror going back and forth from screen to reflection, comparing my physical features to the digital face, adjusting sliders set to individual points on the face. Wider nose, higher cheekbones, square jaw, or to use a referencing program such as PureRef which allowed me to overlay a window with photographs of my face, set at high transparency so I could adjust the Metahumans sliders accordingly (Fig. 1).

Both methods produced inaccuracies. For example, I was unable to recreate the exact bump of the bridge of my nose. I felt a subconscious desire to "fix" or idealise my real features in the virtual copy. As the starting point was built on pre-existing faces it never truly felt like my own. As I began utilising Live Link Face (launched by Unreal in July 2020) I was able to animate my Methuman in real-time with facial animation captured through my iPhone (Fig. 2). Through this interaction I felt I was able to build somewhat of a closer connection with my avatar, it didn't look completely like me but now I could control how it moved and spoke.

When it came to adding a voice I felt anxious. As a trans woman the pitch, tone and resonance of my voice can be a source of being misgendered by others. As my avatar was reading a mixture of words written by myself and poems co-written with a writing style A.I., I chose to add subtle reverb and distortion to better suggest a synthesis between myself and my digital copy. For the final render, I used Speech Graphics which automated the animation process, generating facial expressions, eye movements and synched lips to the prerecorded audio. This added a further layer of the uncanny and a "fakeness" to the Metahuman as there was no contraction or relaxing of my facial muscles and bone structure.

When Epic Games first launched the Metahuman Creator software in February 2021, it used a sculpting tool allowing facial features to be manipulated in real-time

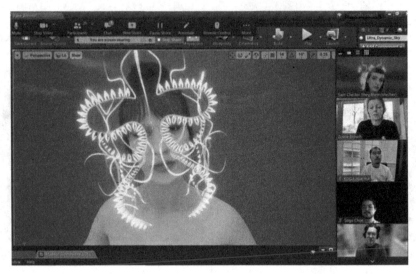

Fig. 2. Screenshot of Zoom session running Live Link Face to animate Metahuman (source: Author's own)

within "physically plausible" limits. The tool allowed users to "blend" between multiple Metahumans, adjusting sliders, emphasising or reducing elements from the blended presets. According to Unreal Engine (2023), the catalogue of faces introduced at the time was "based on pre-existing scans of real people" allowing for "easy" production of realistic digital humans.

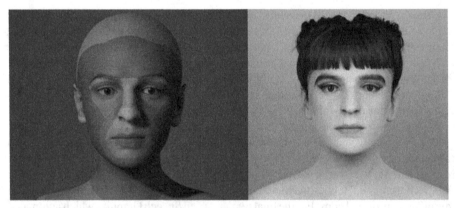

Fig. 3. (left) Screenshot of "Mesh to Metahuman" Scan, (right) Screenshot of "MetaSam" with textures & hair assets (source: Author's own)

More recently, Epic has released a plugin called Mesh to Metahuman, giving people the ability to upload a 3D scan of a face that is rigged as a Metahuman. This is a step forward and one that I used to create my "MetaSelf" (Fig. 3). However it comes with limitations, to fix any issues with the original scan you are given the option of smoothing out any abnormalities in the mesh aligning your original scan more closely to the "perfection" of the ones created in the studio. Secondly, it doesn't retain the textures of the original scan, instead, you have to create a close facsimile from the skin textures provided (Fig. 3).

Fig. 4. Screenshots of 3D models of myself - Agisoft Metashape (source: Author's own)

Further Experimentation: Prior to my designs in the Metahuman Creator I achieved raw energy and intensity with other software.

I attempted to represent my gender identity by taking a series of 360° photographs at various heights and stitching them together in Agisoft Metashape software, to create a 3D model (Fig. 4). It was the first time I had seen myself in 3D space, the process of photogrammetry involves taking multiple overlapping images of a subject, the distance of the camera from each picture is then estimated. This produces a point cloud of data which can be turned into a polygonal mesh, a highly skilled task requiring an expensive camera rig and lighting setup. The process of capturing my image was amateur and led to multiple glitches, seams and ruptures in the model. My face was captured from multiple angles, and my hair and other fine details were matted and fused, these abstract collages of various angles of my face and body to me better reflected my gender identity, an identity that was in a state of flux. This felt closer aligned to my sense of crossing gender binaries, inhabiting a moment in-between spaces, the actual process of having a friend take all the photos required me to stand as still as possible for 5 to 10 min.

As an undergraduate at the RCA, I was able to attend professional workshops, where I took a full body scan using an Artec Eva colour scanner (Fig. 5). This was essentially the same process I had used before but at a much higher fidelity. The 3D model, with fewer glitches, was a copy of my physical likeness. At the time I was still identified as non-binary, which led to the 3D model being customised to remove any obvious sexual characteristics (Fig. 6 and Fig. 7).

3.2 Portraying Hearing Impairment in the Metahuman

Case Study #2: Seb. I trained as a fashion designer and currently conduct research into inclusive and accessible design and technology. I lost my hearing at the age of 18 due to a rare neurological condition and learned to communicate through lipreading, to the

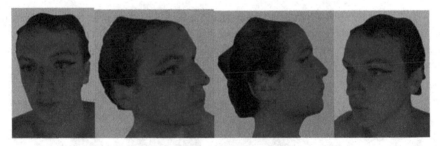

Fig. 5. Screenshots of my head captured with an Artec Eva Scanner (source: Author's own)

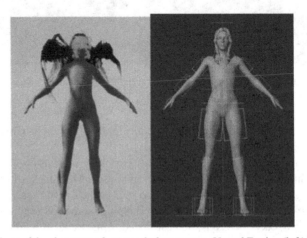

Fig.6. Screenshots of development of my genderless avatar - Unreal Engine (left)/Blender (right) (source: Author's own)

Fig.7. Screenshot of my final film "At_the_Sea's_Edge" - Unreal Engine post (source: Author's own)

point where hearing people rarely notice my disability. Whilst I may be perceived as able-bodied, my experience continues to be that of someone who is profoundly deaf.

I cannot hear my surroundings, and cannot respond to or process the auditory world around me.

During my postgraduate study at the RCA I developed different experiences which translated sound into visual and tactile stimuli. I began by experimenting with three-dimensional digital sculptures created from sound, based on an algorithm utilising pitch, volume, timbre, and other aspects of noise. While these digital sculptures do not allow the viewer to understand the sound in a rational sense, they were the first steps to translating an emotional layer of sound. It was never about the ability to understand sound from a logical perspective, rather from an emotional dimension. The aim was to develop garments which translated sound into tactile stimuli, enabling people who are deaf to experience sound without hearing it.

Aim. The aim of my Metahuman was to explore how my identity as a person who is unable to decipher sound (an invisible disability) might be represented.

Creative Process. I was intentional in my design of my avatar and set out to create a digital facsimile of myself somewhat like a digital mirror, which looked like me, moved and spoke like me. This was my first experience of digital identity creation, having previously experimented with I spent a total of 17 h designing my Metahuman to be my digital twin (Fig. 8) but already failed with the first intention, to create a Metahuman which looks like me.

Fig. 8. Representation of myself in the Metahuman (source: Author's own)

While the software offered a wide range of facial features and skin complexions, choice of hair and eye colour and whilst I got close to it at points, it never looked right or felt right. The first reason for this is that the sculpting tools in the creator were quite slow and clunky to respond, and while it offers a lot of possible changes, it is also quite restrictive in what the Avatar can look like.

It took me some time to figure out the second reason: while the Metahuman Creator's realistic digital human characters are based on 80 pre-existing scans of real people, it aestheticises everything that is created. There was no way of creating bodies that stray from the hegemonic norms. This was quite demotivating and I abandoned my original plan to create my digital twin. I changed my objective to explore if different

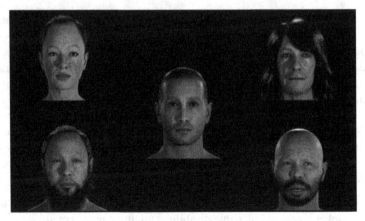

Fig. 9. A Collective of 5 Perfect Imperfections (source: Author's own)

disabled identities could be created in the Metahuman Creator. I customised the physical characteristics of different preset avatars, facial features, hair, hair colour, and eyes to represent the features of people with visible disabilities. As there was no way for the Metahuman Creator to design an avatar with missing limbs or other body impairments, I focused on facial features. But while trying to break away from the able- bodied avatars, changing features like forehead width, eye distance, angle, and size, (Fig. 9) the limits of what changes can be made, made it impossible to represent the features of people with disabilities. Even worse, I found myself facing the possibility of perpetuating harmful stereotypes around disability, race and gender. I needed more fine-grained customisation to remove the disabling barriers of the software.

For the final outcome of my Metahuman project, I decided to go in a different direction. I arranged my five Metahumans, which show signs of inherent aestheticization into a "collective" of perfect imperfection. This collective recites a message (Fig. 10) to the viewer about the viewers' unworthiness to be in the Metaverse.

> We are human
> Perfect
> Complete
> You don't belong here
> You are damaged
> Wrong
> Unworthy
> People like you
> Don't exist in the metaverse
> There is no place
> For Imperfection
> No diversity
> No inclusivity
> We don't need it
> You don't exist

Fig. 10. The descriptive text the Collective of 5 Perfect Imperfections recite (source: Author's own)

In order to represent my experience of not hearing the world, the Metahumans move their lips but no sound is audible. The able-bodied viewer can only understand through the subtitles at the bottom of the video.

4 Findings and Analysis

Diverse user groups. The findings show a need to include visible and invisible disabilities, trans and gender non-conforming individuals within the scanned catalogue of digital humans supplied by Epic. This would open up the possibility of blending faces and features outside default options. Greater customisation would enable (i) trans users to create a feminine body with broader shoulders, narrower hips, and smaller breasts (ii) visibly disabled users to create disability-related avatar features and (iii) invisibly disabled users to choose gestures (signing) as part of the Metahumans' preset tools (Figs. 11 and 12).

Fig. 11. Examples of Metahuman expression presets (source: Author's own)

Fig. 12. Examples of Metahuman movement presets (source: Author's own)

Customisable Options. An expanded library of customisable makeup colours and styles is needed. Makeup and other cosmetics are not merely mundane rituals but, as Erickson-Schroth (2014: 74) argues, "profound acts of self-care, affirmation and resilience", vital for building confidence in the trans community. Support for original, as opposed to pre-existing, textures in the Mesh to Metahuman feature would allow a person's unique skin type, texture, and distinguishing markings to be applied directly to the Metahuman.

These findings are congruent with the experience of gaming communities interviewed by Kosceisza (2023), van Aller (2018) and Morgan (2020). Kosceisza (2023:10) argued that trans-feminine players made "strategic choices", by "pitching [their] voice" higher, to avoid harassment from other players. The same anxiety is evident in Sam's response, where she is concerned that her disembodied voice will be "witnessed" as masculine due to her natural intonation, resonance, articulation and vocal quality. The sociologist, Aaron Devor (2004), in his model of "Transsexual Identity Formation", describes humans as social beings with a deep need to be 'witnessed' by others for who we are. Hence, when transgender individuals are validated by dispassionate, i.e., impartial, others in ways that accurately conforms to their own sense of self, their identities are powerfully reinforced. Conversely, a gender dichotomised virtual world has the effect of inflicting psychological distress.

Whilst face and body controls are not gendered within the Metahuman Creator, the platform's online literature lists body proportions as either "feminine" or "masculine body types" with measurements "representative" of men and women (Fig. 13).

The following heights are representative of the feminine and masculine body types:

Height	Feminine Type	Masculine Type
Short	4'11" (1.49m)	5'5" (1.65m)
Average	5'3" (1.60m)	5'8" (1.72m)
Tall	5'6" (1.675m)	5'11" (1.80m)

Fig. 13. Body   proportions: representative   of   male   and   female   body   types (source: metahuman-creator-body-controls 2023)

The tallest Feminine Type is measured at $5'6''$, which means that Sam's height of $5'11''$ lists under Tall Masculine, a muscular athletic physique unlike her own. As van Aller argues (2018: 15) when deciding "options for bodies, facial features, hairstyles, clothing, voice pitch, and pronouns" (2018: 15), it is necessary to include trans and gender non-conforming users.

The inability to create avatars with invisible markers of disability are currently based on what is physically plausible for able-bodied people. As Harper argues (2020: 276) it may be that we need a digital human configurator where the disabled body is the modular core, as we are prone to "valorise able-bodied norms of inclusion" as the qualification for citizenship (Mitchell & Snyder 2015: 12). We must design for a future Metaverse that is not merely a mirror of the physical present, but halts the cycle of "mutant bodies" being marginalised on the edges of power and refused their fundamental human right to exist.

5   Conclusion

5.1   Moving Beyond the Present Impasse

At the beginning of this perspective paper, we set out to answer the following research question:

To what extent does the Metahuman embody "human dignity" in the representation of self in the virtual world?

We recognise that whilst the United Nations' Charter (1945) defines Human Dignity as "the worth of the human person", in the participant studies cited, marginalised identities are not afforded this human right.

The limitations we have identified in the Metahuman Creator are symptomatic of the challenges to be addressed in the emergent Metaverse. As Anne Balsamo (1996:131) predicted, we have seen "old identities … continue to be more comfortable, and thus more frequently reproduced" in virtual spaces. We argue that tools for designing identity online must strive for a Pluriverse "in which diverse hopes can be sown, multiple opportunities cultivated, and a plurality of meaningful lives achieved" (Gaard 2017). Pluriversality as defined by Mignolo (2018) views the world as an "interconnected diversity" rather than a unified totality, so that in order for it to make sense it rejects Christian, Liberal, Marxist models of universal ownership in favour of indigenous and decolonised models such as the Zapatistas who create "Queremos un mundo donde quepan muchos mundos", a world where many worlds fit (Rogers, 2021).

In Designs for the Pluriverse, Escobar (2017) describes "design autonomy" as a set of "tools, interactions, contexts, and languages" that empower all forms of ontological world building. This expansion of meaning-making through pluriversality offers an opportunity to develop and extend our definitions of human dignity in relation to the existing definition of the Metaverse.

In the next phase of our research, we intend to work with both disabled and trans communities to gather data that is representative of the multiplicity of experiences within these groups. We will reach out to organisations such as WITCih (The Women in Technologies Creative Industries Hub) Digi-GXL (global group of womxn, intersex, trans, non-binary people specialising in digital design) and the Helen Hamlyn Centre for Design at the Royal College of Art to gather respondents who are actively engaging in digital technologies. We hope to collaborate with digital human configurators such as Epic in designing new tools, practices and processes that will enable an authentic representation of disabled and gender fluid bodies.

Acknowledgments. We would like to thank Sallyann Houghton, Epic Games and Speech Graphics for supporting the project and to all of the participants from the Royal College of Art's School of Design for engaging in this study. Thank you also to Kam Roofi and Malcolm Pate for their invaluable assistance with technology.

References

Ball, M.: Framework for the Metaverse (2021). https://www.matthewball.vc/all/forwardtothemet averseprimer. Accessed 15 Sept 2023

Balsamo, A.: Technologies of the Gendered Body: Reading Cyborg Women. Duke University, North Carolina (1996)

Barad, K.: Posthumanist performativity: toward an understanding of how matter comes to matter. In: Åsberg, C., Braidotti, R. (eds.) A Feminist Companion to the Posthumanities, pp. 223–239. Springer, Cham (2018). https://doi.org/10.1007/978-3-319-62140-1_19

Derry, T.K., Williams, T.I.: A Short History of Technology: From the Earliest Times to A.D. 1900. Dover Publications, New York (1993)

Devor, A.H.: Witnessing and mirroring: a fourteen stage model of transsexual identity formation. Gay Lesbian Mental Health. **8**, 41–67 (2004)

Dewey, J.: Philosophy and Civilisation: The Quest For Certainty Individualism Old and New. Minton Bach & Co, New York (1927)

Erickson-Schroth, L.: Trans Bodies Trans Selves: A Resource for the Transgender Community. Oxford University Press, Oxford (2014)

Escobar, A.: Designs for the Pluriverse: Radical Interdependence, Autonomy, and the Making of Worlds. Duke University Press, Durham (2017)

Ferrando, F.: Philosophical Posthumanism. Bloomsbury, London (2019)

Floridi, L.: Metaverse: a matter of experience. Philosophy & Technology (2022). https://ssrn.com/abstract=4121411. Accessed 16 July 2023

Gaard, G.: Power of the Pluriverse (2017). https://medium.com/microsoft-design/power-of-the-pluriverse-c3592f5eca76. Accessed 15 July 2023

Harper, T.: Endowed by Their Creator: Digital Games, Avatar Creation, and Fat Bodies. Fat Stud. **9**(3), 259–280 (2020). https://doi.org/10.1080/21604851.2019.1647022. Accessed 20 Sept 2023

Harari, N.Y.: Homo Deus: A Brief History of Tomorrow. Vintage, London (2017)

Mignolo, W.: On Pluriversality and Multipolarity, Foreword ix, in Pluriverse: The Geopolitics of Knowledge. Edited by Bernd Reiter. Duke University Press, Durham, USA (2018)

Morgan, H., O'Donovan, A., Almeida, R., Lin, A., Perry, Y.: The role of the avatar in gaming for trans and gender diverse young people. Int. J. Environ. Res. Public Health **17**(22), 8617 (2020). https://doi.org/10.3390/ijerph17228617

Nowak, K.L., Fox, J.: Avatars and computer-mediated communication: a review of the definitions, uses, and effects of digital representations. Rev. Commun. Res. **6**, 30–53 (2018)

Papacharissi, Z. (ed.): A Networked Self: Identity, Community, and Culture on Social Network Sites. Routledge, London (2010)

Radanliev, P., Roure, D., Novitzky, P., Sluganovic, I.: Accessibility and Inclusiveness of New Information and Communication Technologies for Disabled Users and Content Creators in the Metaverse. https://doi.org/10.2139/ssrn.4528363. Accessed 20 Oct 2023

Rogers, E.W.: The Metaverse must be a Plurivers, 24 December 2021. http://ericwycoffrogers.com/writings/2021/12/24/the-metaverse-must-be-a-pluriverse. [Accessed 16/07/23]

Schachter, O.: Human dignity as a normative concept. Am. J. Int. Law **77**(4), 848–854 (1983)

Schultze, U.: The Avatar as Sociomaterial Entanglement: A Performative Perspective on Identity, Agency and World-Making in Virtual Worlds, AIS Electronic Library (AISeL) (2011). https://core.ac.uk/works/17801639. Accessed 20 Oct 2023

United Nations: United Nations Charter: Preamble (1945). https://www.un.org/en/about-us/un-charter/preamble. Accessed 14 July 2023

Unreal   Engine.   https://www.unrealengine.com/en-US/metahuman?utm_source=GoogleSea rch&utm_medium=Performance&utm_campaign=19729682794&utm_id=146146760853& utm_term=metahuman%20creator&utm_content=649132338851www.unrealengine.com/. Accessed 26 Oct 2023

USTSExecutiveSummary. https://transequality.org/sites/default/files/docs/usts/USTS-Executive-Summary-Dec17.pdf [Accessed 20/07/23]

Vasalou, A., Joinson, A., Banziger, T., Goldie, P., Pitt, J.: Avatars in social media: balancing accuracy, playfulness and embodied messages. Int. J. Human-Comput. Stud. **66**(11), 801–811 (2008). https://doi.org/10.1016/j.ijhcs.2008.08.002

Wilke, A., Savransky, M., Rosengarten, M.: Speculative Research: The Lure of Possible Futures. Routledge, Oxon (2017)

World Health Organization: Disability. https://www.who.int/news-room/fact-sheets/detail/disabi lity-and-health. Accessed 30 July 2023

Zallio, M., Clarkson, P.J.: Designing the metaverse: A study on inclusion, diversity, equity, acces-sibility and safety for digital immersive environments. Telem. Inform. **75**, 101909 (2022). https://doi.org/10.1016/j.tele.2022.101909

Using Generative Models to Create
a Visual Description of Climate Change

Felipe Santana Dias[1]([✉]) [iD], Artemis Moroni[2] [iD], and Helio Pedrini[1] [iD]

[1] Institute of Computing, University of Campinas, Campinas, SP 13083-852, Brazil
felipesantanadias@gmail.com
[2] GAIA Senses, Renato Archer Information Technology Center,
Campinas, SP 13069-901, Brazil

Abstract. The discrepancy between the rapid dissemination of information and its effective communication underlies the phenomenon of scientific denialism. Given the rapid strides in AI generative models, this project explores the domain of knowledge visualization to portray weather data through a visually captivating representation of Rio de Janeiro's climate evolution up to the year 2100. The use of prompt engineering over climate models has yielded promising outcomes in image generation, yet challenges remain in ensuring deterministic accuracy in image construction.

Keywords: Generative Models · Knowledge Visualization · Climate Change

1 Introduction

In a survey conducted by Yale in 2021, 47% of Americans said they do not believe that global warming will harm them personally [21]. Despite scientific advancements that highlight and substantiate ongoing climate changes, this statistic reveals the significant gap between the access of information and the acceptance among the general population.

To combat scientific denialism, the World Economic Forum introduced the acronym SUCCES based on the book "Made to Stick" by Chip and Dan Heath [20]. According to this framework, it is crucial to share information in a **simple, unexpected, credible** manner, presenting the data in a **concrete** way, using analogies and metaphors to connect with the **emotional** aspects of the audience, shaping scientific knowledge into compelling **stories** [8].

Thus, when it comes to climate change, creating accurate visual representations of abstract information becomes interesting in fostering a stronger connection and raising awareness of the magnitude of this problem.

Therefore, this paper presents the construction of a cultural product, i.e., an audiovisual work that has the purpose of large-scale consumption, presenting

Supported by GAIA Senses - Renato Archer Information Technology Center, Campinas, SP, Brazil, 13069-901.

A. L. Brooks (Ed.): ArtsIT 2023, LNICST 564, pp. 202–212, 2024.
https://doi.org/10.1007/978-3-031-55319-6_14

climate data and scientific statements in a concrete way with strong emotional appeal aiming to enhance acceptability and engagement among the audience.

Based on models of weather factors provided by CMIP6, a initiative of the World Climate Research Programme (WCRP) [16], climate factors up to the year 2100 will be visualized within a unified artistically pictorial landscape based on Guanabara Bay, Rio de Janeiro - one of the most susceptible areas to the impacts of climate shifts [3]. This project named Vigilante, in reference to residential security personnel, seeks to allow viewers to envision a possible future under new conditions.

This study aligns with the field of Knowledge Visualization (KV), a relatively young discipline that explores "all (interactive) graphic means that can be used to develop or convey insights, experiences, methods, or skills" [13–15].

Historically, the evolution of graphic information can be traced from the Middle Age manuscript culture in Europe to contemporary computer-based visualization [31]. With the rapid and significant progress of generative AI models, the synthesis capabilities of these algorithms combined with a systematic data-driven approach can create alternative means to represent abstract information, advocating a novel mode of engaging with data through pictorial images that ensures greater comprehensibility for the audience.

Prominent examples of this models includes DALL-E [30], Mid Journey [27], and Stable Diffusion [33], which are capable of generating complex synthetic images based on descriptive prompts. Also, applications such as Deforum [9] and Kaiber.ai [23] enable the creation of videos by concatenating multiple images from generative models.

In light of these considerations, we aim to address the following research question: "Is it effective to use current generative models based on prompt engineering to represent deterministic data?"

Our contributions are as follows:

(i) Introduce the use of pictorial images for information representation.
(ii) Developing a pipeline for systematic and data-driven utilization of generative models as tools for innovative information visualization.
(iii) Present a graphical representation depicting the climate transformation in Rio de Janeiro over the upcoming century.

2   Literature Review

The use of computers to generate creative works is an area that has been studied and validated in academia [28]. When it comes to knowledge visualization, the literature suggests that generated images should provide assistance for reasoning, reflection, and the exploration of connections in new ways, in order to facilitate new discoveries based on shared insights [13]. It is also understood that pictorial representations reveal significant aspects of the history of visual culture and knowledge [12].

Regarding art made with generative models that address environmental issues, some artists have created visual representations of climate change and its consequences in landscapes more sensitive to global warming such as glaciers and coral reefs with the aim of raising awareness about the cause [4,19].

3   Methodology

Drawing on the concept that offering comprehensible and accessible data can enhance individuals' awareness of climate changes and promote action for the cause [32], our poetical proposal is to produce a video where each frame reflects the conditions of climate variables predicted during sunset, where the colors conveying the tranquility and melancholy of the end of the day while also reflects the uncertainty of the future.

Observing the engagement in discussions about climate change on social media, 56% of users from Generation Z (born after 1996) stated that they interacted with this topic in the past week, compared to 44% of users from other generations, namely Millennials (born between 1981–1996), Gen X (born between 1965–1980), Boomers (born between 1946–1964), and older [35]. Also, 69% of Generation Z social media users stated that climate change content made them feel anxious about the future, with one of the causes being the dissatisfaction with the insufficient amount of current actions being taken [35].

As "the 'aura' of the digital object is fundamental to how it is received by its audiences" [22], recognizing the feelings of anxiety and dissatisfaction can be interesting in driving action in favor of the environment, developing a project that provides foundation and tools for Generation Z to express their opinions and viewpoints beyond their own generation [18]. An overview of the project's pipeline can be seen in Fig. 1.

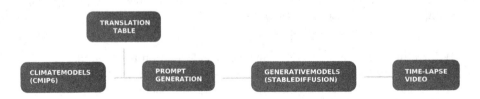

Fig. 1. Pipeline to create images based on satellite data.

According to the Global Climate Observing System (GCOS), there are essential climate variables (as listed in Table 1) that help us understand the behavior of the global warming [5].

Utilizing the surface atmosphere variables models from CMIP6 Amon table with monthly frequency (*mon*), with geographic coordinates −23, −42, the data spanned from 2015 to 2100 representing a fossil fuel-based developed world focused on rapid global economic expansion (SSP5 scenario) [2].

As observed in Fig. 2, the increasing trend in average temperature, coupled with higher levels of precipitation, creates a conducive environment for flooding and landslides [3,24]. The rise in solar radiation and atmospheric pressure further exacerbates urban heat islands. The possibility of abrupt pressure changes also leads to an increase in the occurrence of storms in the region [17].

Table 1. Climate variables and models.

GCOS Variable	CMIP6 Model [6]	Prompt Variable
Precipitation	Precipitation (pr)	Precipitation, Sky conditions
Pressure	Sea Level Pressure (psl)	Sky conditions
Radiation Budget	Surface Downwelling Clear-Sky Longwave Radiation (rldscs)	Sea level
Temperature	Near-Surface Air Temperature (tas)	Windchill
Wind Speed	Near-Surface Wind Speed (sfcWind)	Windchill, Wind

Prompt Engineering. Although current image generative models are primarily developed for natural language understanding, achieving desired results hinges on an individual's ability to accurately describe the object [25,29]. Several guidelines are proposed for the methodological production of high-quality images, as the following template [29]:

[Medium] [Subject] [Artist(s)] [Details] [Image repository support]

In order to represent deterministic information, we created prompts from the model data. Numerical values were translated into descriptive English terms, aligned with meteorological language and human climate perception (Fig. 3). For more complex scenarios involving multiple factors, translations draw from researcher-described situations rather than raw data abstraction. For artistic aims, other prompt engineering criteria rely on frequently employed keywords for visual descriptions as represented in Table 2.

The use of prompts to represent a same location with specific climate variations can lead to out-of-context elements in the image defined as hallucinations [26].

When comparing real landscapes that have identical climate data, the visual composition is not strictly the same between the environments, possessing a randomness that can be likened to the hallucinate process of generative models. However, it can challenge the goal of assigning deterministic meaning and maintain a visually coherent sequence while accurately depicting the imposed changes.

Temporal Connection. Once the method for generating high-quality images that visually represent the data was established, it became necessary to create images that maintain the connection between subsequent frames in order to present the predictions as a video.

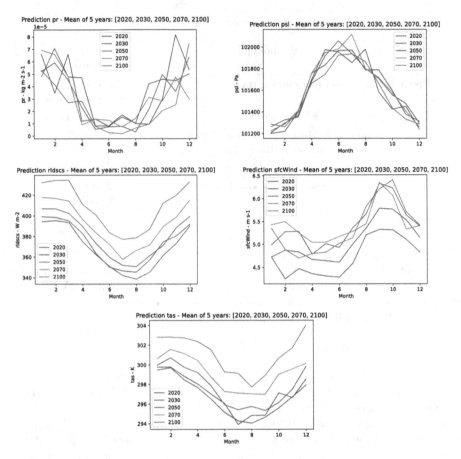

Fig. 2. Precipitation, Pressure, Radiation Budget, Wind Speed, and Temperature distribution over years.

To enhance control over hallucination and temporal transitions, a viable strategy to stimulate or repress this effect involves fine-tuning the interplay between the image_strength and cfg_scale parameters. These parameters determine the degree of impact the initial image wields on the diffusion process and how closely it adheres to the prompt text, respectively.

In this study, it was experimented the creation of videos using time-lapse approach, where each individually frame was generated using Stable Diffusion API v1.5 or xl-beta-v2.2.2 based on the previous image generated with random seed, parameter that ensures reproducibility and control over the generated outputs. This accumulation factor of images enables the representation of subtle climate changes that occur over time, such as rising sea levels, but it also accentuates certain aspects of the image that distort the environment.

Table 2. Keywords to create prompt variations [7,34].

Medium	Subject	Artist	Details	Image Repository Support
A photograph of	Guanabara Bay,	VSCO, Pinterest,	figurative style = renaissance style, hyperrealism,	Quality Booster = 4k, award-winning photograph,
	{Precipitation},		{Time of the day},	90s
	{Sky conditions},			
	{Sea level},			
	{Windchill},			
	{Wind},			

Variable	Values
{Precipitation}	'dry day, clear sky', 'moderate rain, few clouds', 'storm, cloudy day', 'heavy storm, thunders, lightning'.
{Sky conditions}	'vibrant', 'overcast', 'melancholic atmosphere', 'lightning, darker atmosphere'.
{Sea level}	'calm sea', 'sea waves invading the sand', 'sea waves invading the sidewalk', 'sea invading the street and reaching houses', 'sea taking over the city'.
{Windchill}	'nice weather, lush vegetation', 'hot weather', 'dry vegetation', 'arid landscape'.
{Wind}	'calm wind', 'fresh wind', 'soft wind', 'strong wind, wind vegetation', 'hurricane, destroyed vegetation'.
{Time of the day}	'sunrise', 'morning', 'golden hour', 'sunset', 'night'.

4   Results

In the context of controlled generative image production, specific configurations play a pivotal role in shaping the visual outcomes. The video featured in [10] showcases the results obtained.

When the image_strength is equal to zero and the seed value is fixed, the generated images do not exhibit visual correlation among themselves when subtly altering the prompt. Conversely, when alternating between different prompts, consistently similar images are generated in accordance with the prompts used (Fig. 4).

For image_strength parameter with values above 0.75, the changes in the image are too subtle, not being able to distinguish the individual data from each frame. Consequently, after multiple iterations, the image is distorted by noise, as observed in Fig. 5.

When the image_strength parameter is set from 0.7 (Fig. 6), a temporal correlation is established while the generated images assume the characteristics described by the prompt. However, with an increasing number of iterations, the

Fig. 3. Prompt modifications on the field *subject* to generate images with different seeds: (Fig. 1) clear sky, lush vegetation, few boats in the water, calm sea; (Figs. 2 and 3) cloudy sky, destroyed vegetation, drown boats in the water, lightning, stormy sea; (Fig. 4) clear sky, vegetation covered in snow, frozen water, sunset lights, calm sea.

Fig. 4. Stable Diffusion, text-to-image, seed = 2792954258.

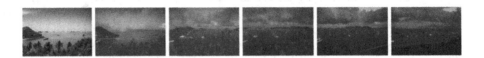

Fig. 5. Stable Diffusion, image-to-image, image strength = 0.75.

Fig. 6. Stable Diffusion, image-to-image, image strength = 0.7.

Fig. 7. Stable Diffusion, Image-to-Image, image strength = 0.6.

Fig. 8. Stable Diffusion, image-to-image, multiple frame transition.

model tends to hallucinate, generating random elements that do not relate to the environment, which can be desirable in some contexts, such as the artistic field. In one of the experiments, during testing with an image_strength of 0.6, the number of boats exponentially increased occupying the scene completely (Fig. 7).

From then on, we adopted a combination of the parameters with random seed while incorporating multiple frames for transitions. Following a series of iterative experiments, an optimal configuration materialized: a three-frames transition involving two distinct model versions [xl-beta v2.2.2, v1.5, v1.5] accompanied by image_strength values [0.8, 0.7, 0.7] and cfg_score values [3, 6, 9]. The outcome entailed the production of images exhibiting heightened thematic interconnectedness, coupled with an improved capacity to capture nuanced variations in individual climates (Fig. 8).

5   Discussion

The application of generative models in crafting data-driven knowledge visualizations, particularly for individual images, has demonstrated a remarkable potential. In cases where images are generated without previous frame, a resemblance to actual landscapes emerges, capturing the essence of the visual representation found in data. Nevertheless, the introduction of temporal connectivity presents challenges.

While the inclusion of previous images in the generation process enables consistency between frames, it often sacrifices the accurate portrayal of individual image characteristics, being traded off due to unintended hallucinations from the essence of the original landscape when previous images are not heavily included.

The prompt engineering approach, while contributing to the accessibility of the models, has inherent linguistic limitations. Climatic description in natural language is location-sensitive. Describing a Brazilian city in English introduces linguistic and visual biases due to translation.

Due to the process of describing the data in natural language, when processed by generative models, they may not interpret the linguistic nuances exactly as intended. Consequently, the created visual representations may exhibit significant bias stemming from the author's perception of the acquired information.

Moreover, these models provide a gateway to the realm of imagination, rendering them a valuable asset in scientific communication. Their use in academia allows for visual communication of findings, bridging the gap between complexity and accessibility. However, despite these advancements, deterministic accuracy remains an elusive aspiration through this strategy.

6   Conclusions and Future Work

The Vigilante project generated interesting results by observing the progression of climatic events in the city of Rio de Janeiro. However, due to the hallucinatory capabilities of generative models, they currently lack the ability to assume a deterministic nature.

Deepfake techniques present a promising means to gain more control over hallucination during image generation [36]. Implementing elements of control can further ensure that the outcomes of this work do not distort potential realities of the near future. In conjunction with these changes, the establishment of objective evaluation metrics would allow the assessment of multiple models, climate variables and keyword combinations, automating the evaluation of generated results in comparison to desired outcomes.

To augment realism, training generative models on image databases specifically tailored for each ecosystem is essential. There is also a growing interest in developing mask technologies capable of maintaining overall coherence, especially when generating images that depict particular scenarios under diverse weather conditions.

Research suggests that, regardless of their technical prowess, the mere awareness that a creation is computer-generated can diminish its artistic value [1]. Hence, gauging the emotional resonance of the generated videos and evaluating the impact of such visualizations on human understanding and perception are crucial for effective communication. Undertaking field research to acquire a comprehensive grasp of the audience's perspectives and concerns will enable more tailored visualizations, fostering a deeper connection with viewers.

Lastly, an exemplar of the transformation of the acquired content into a cultural product can be accessed at [11].

Acknowledgements. I extend my gratitude to Dr. Priscila Coltri for her valuable comments and guidance on climate resources, to Matheus Alves for his assistance in data acquisition and processing, to Davide Romano for his collaboration with references, and to Michael Al-Hussein for providing his footage of Rio de Janeiro.

References

1. Agudo, U., Arrese, M., Liberal, K.G., Matute, H.: Assessing emotion and sensitivity of AI artwork. Front. Psychol. **13**, 879088 (2022)
2. Riahi, K., et al.: The shared socioeconomic pathways and their energy, land use, and greenhouse gas emissions implications: an overview. Global Environ. Change **42**, 153–168 (2017). https://doi.org/10.1016/j.gloenvcha.2016.05.009, https://www.sciencedirect.com/science/article/pii/S0959378016300681
3. Alisson, E.: Sea levels along the Brazilian coast are expected to rise in coming decades. FAPESP Agency (2023). https://agencia.fapesp.br/sea-levels-along-the-brazilian-coast-are-expected-to-rise-in-coming-decades/25560/#:~:text=In %20the%20city%20of%20Rio1963%2D2011%20at%20a%2095%25
4. Anadol, R.: Artificial Realities: Coral (2022). https://refikanadol.com/works/artificial-realities-coral/. Accessed 30 June 2023
5. Bojinski, S., Verstraete, M., Peterson, T.C., Richter, C., Simmons, A., Zemp, M.: The concept of essential climate variables in support of climate research, applications, and policy. Bull. Am. Meteor. Soc. **95**(9), 1431–1443 (2014)
6. Boucher, O., et al.: IPSL IPSL-CM6A-LR model output prepared for CMIP6 ScenarioMIP ssp585 (2019). https://doi.org/10.22033/ESGF/CMIP6.5271
7. Cascallar-Fuentes, A., Ramos-Soto, A., Bugarín, A.: Meta-heuristics for generation of linguistic descriptions of weather data: experimental comparison of two approaches. Fuzzy Sets Syst. **443**, 173–202 (2022)
8. Cook, J.: How to counter science denial. World Economic Forum, June 2015. https://www.weforum.org/agenda/2015/06/how-to-counter-science-denial/
9. Deforum-Art/Deforum-Stable-Diffusion (2023). https://github.com/deforum-art/deforum-stable-diffusion. Accessed 27 June 2023
10. Dias, F.: Vigilante|Academic Presentation (2023). https://youtu.be/iD6-e3jIem0. Accessed 18 Aug 2023
11. Dias, F.: Vigilante|Official Video (2023). https://youtu.be/IH8-VO8xo4g. Accessed 18 Aug 2023
12. Drucker, J.: Graphesis: Visual Forms of Knowledge Production, vol. 6. Harvard University Press, Cambridge (2014)
13. Eppler, M.J.: What is an effective knowledge visualization? Insights from a review of seminal concepts. In: 15th International Conference on Information Visualisation, pp. 349–354. IEEE (2011)
14. Eppler, M.J., Burkhard, R.A.: Knowledge visualization: towards a new discipline and its fields of application. Università della Svizzera Italiana, Technical report (2004)
15. Eppler, M.J., Burkhard, R.A.: Visual representations in knowledge management: framework and cases. J. Knowl. Manag. **11**(4), 112–122 (2007)
16. Eyring, V., et al.: Overview of the coupled model intercomparison project phase 6 (cmip6) experimental design and organization. Geosci. Model Dev. **9**(5), 1937–1958 (2016). https://doi.org/10.5194/gmd-9-1937-2016, https://gmd.copernicus.org/articles/9/1937/2016/
17. de Faria Peres, L., de Lucena, A.J., Rotunno Filho, O.C., de Almeida França, J.R.: The urban heat island in Rio de Janeiro, Brazil, in the last 30 years using remote sensing data. Int. J. Appl. Earth Obs. Geoinf. **64**, 104–116 (2018)
18. Fromm, J., Garton, C.: Marketing to millennials: Reach the largest and most influential generation of consumers ever. Amacom (2013)

19. Harris, G.: Artist Refik Anadol brings climate crisis to a scorching Basel with AI-generated glacier installation. The Art Newspaper, June 2023. https://www.theartnewspaper.com/2023/06/13/artist-refik-anadol-brings-climate-crisis-to-a-scorching-basel-with-ai-generated-glacier-installation

20. Heath, C., Heath, D.: Made to stick: Why some ideas survive and others die. Random House (2007)

21. Howe, P.D., Mildenberger, M., Marlon, J.R., Leiserowitz, A.: Geographic variation in opinions on climate change at state and local scales in the USA. Nat. Clim. Chang. 5(6), 596–603 (2015)

22. Jeffrey, S.: Challenging heritage visualisation: beauty, aura and democratisation. Open Archaeol. 1(1), 1–9 (2015)

23. Kaiber.ai. https://kaiber.ai. Accessed 27 June 2023

24. Kulp, S.A., Strauss, B.H.: New elevation data triple estimates of global vulnerability to sea-level rise and coastal flooding. Nat. Commun. 10(1), 1–12 (2019)

25. Liu, V., Chilton, L.B.: Design guidelines for prompt engineering text-to-image generative models. In: CHI Conference on Human Factors in Computing Systems, pp. 1–23 (2022)

26. Maynez, J., Narayan, S., Bohnet, B., McDonald, R.: On faithfulness and factuality in abstractive summarization. arXiv preprint arXiv:2005.00661 (2020)

27. Midjourney: Midjourney. https://www.midjourney.com/home/callbackUrl=%2Fapp%2F

28. Moroni, A., Zuben, F.V., Manzolli, J.: Ar T bitration: human-machine interaction in artistic domains. Leonardo 35(2), 185–188 (2002)

29. Oppenlaender, J.: A Taxonomy of Prompt Modifiers for Text-to-Image Generation. arXiv preprint arXiv:2204.13988 (2022)

30. Ramesh, A., et al.: DALL-E 2. https://openai.com/product/dall-e-2

31. Rendgen, S.: History of Information Graphics. Taschen, Köln, [2019] edn. (2019)

32. Shneiderman, B.: The eyes have it: a task by data type taxonomy for information visualizations. In: IEEE Symposium on Visual Languages, pp. 336–343. IEEE (1996)

33. Stable Diffusion: Stable Diffusion Online. https://stablediffusionweb.com/

34. Stewart, A.E.: Linguistic dimensions of weather and climate perception. Int. J. Biometeorol. 52, 57–67 (2007)

35. Tyson, A., Kennedy, B., Funk, C.: Gen Z, Millennials stand out for climate change activism, social media engagement with issue. Pew Res. Center 26, 1–100 (2021)

36. Yadav, D., Salmani, S.: DeeFake: a survey on facial forgery technique using generative adversarial network. In: International Conference on Intelligent Computing and Control Systems, pp. 852–857. IEEE (2019)

Author Index

© ICST Institute for Computer Sciences, Social Informatics and Telecommunications Engineering 2024
Published by Springer Nature Switzerland AG 2024. All Rights Reserved
A. L. Brooks (Ed.): ArtsIT 2023, LNICST 564, pp. 213–214, 2024.
https://doi.org/10.1007/978-3-031-55319-6

Printed in the United States
by Baker & Taylor Publisher Services